Gardner Denver

150 Years of Industrial Innovation

GARDNER DENVER

1859 2009

SESQUICENTENNIAL

150 Years of Industrial Innovation

THE
DONNING COMPANY
PUBLISHERS

The Donning Company Publishers
184 Business Park Drive, Suite 206
Virginia Beach, VA 23462

Steve Mull, General Manager
Barbara Buchanan, Office Manager
Pamela Koch, Senior Editor
Amanda Dawn Guilmain, Graphic Designer
Derek Eley, Imaging Artist
Debbie Dowell, Project Research Coordinator
Tonya Hannink, Marketing Specialist
Pamela Engelhard, Marketing Advisor

G. Bradley Martin, Project Director

Library of Congress Cataloging-in-Publication Data

Gardner Denver : 150 years of industrial innovation
p. cm.
ISBN 978-1-57864-567-1 (hard cover : alk. paper)
1. Gardner-Denver Company—History. 2. Industries—United States—Illinois. 3. Electric apparatus and appliances—United States—Illinois. I. Donning Company Publishers.
HC107.I3B38 2009
338.7'62150973—dc22
2009012707

Printed in the United States of America at Walsworth Publishing Company

Table of Contents

Presidents of Gardner Denver Inc.

Robert W. Gardner
1859–1905

J. W. Gardner
1905–1927

W. H. Leonard
1927–1935

H. G. Myers
1935–1947

Edgar F. Schaefer
1947–1954

Cedric H. Rieman
1963–1973

Gifford Leece
1954–1963

Lynn L. Leigh
1973–1979

Ross J. Centanni
1994–2008

CHAPTER 1

The Gardner Flyball Governor

Quincy was hopping that middle week of October in 1858. The bustling port city on the Mississippi River midway between St. Louis and Keokuk was playing host to what was perhaps the defining political movement of antebellum America. Stephen A. Douglas, the incumbent U.S. senator from Illinois, was locked in a tight re-election race with Abraham Lincoln, a political neophyte who combined a backwoods persona with a solid record as one of the Midwest's most accomplished railroad lawyers.

Douglas and Lincoln had spent much of the summer and fall of 1858 debating the issues up and down the length of Illinois. In the days before radio and television, attack advertisements, and robocalls to voters, campaigning was an intensely personal experience. The best way for Americans to get to know the candidates was up close and personal. And at a time when rhetoric was a valued attribute for anyone aspiring to political office, the formal debate was the preferred vehicle to reach the greatest number of voters at one time. In 1858, the debate over who was best suited to represent the people of Illinois in the U.S. Senate was simple and stark. Slavery had divided America since at least the 1830s, and politicians and statesmen as accomplished as Henry Clay, Daniel Webster, and John C. Calhoun had devoted their careers to either the defense of or an attack against what many called "the peculiar institution."

Now, in mid-October 1858, the debate about slavery had come to Quincy. Both Douglas and Lincoln arrived in Quincy by train, the Senator the evening before, on Tuesday, October 12, and Lincoln at nine o'clock that Wednesday morning. Quincy was the sixth of seven debates between the two held in October 1858. Lincoln and the "Little Giant," as Douglas was known, would board a steamboat in Quincy the day following the debate for a final debate on Friday, October 15, downriver in Alton, Illinois.

Bas-relief of Lincoln-Douglas debate in Washington Park, Quincy, Illinois.
Courtesy of the Historical Society of Quincy and Adams County

When Lincoln arrived in Quincy that Wednesday morning for the 1:30 p.m. debate, more than twelve thousand people awaited the two candidates in Washington Park, nestled three blocks up from the wharves lining the Quincy riverfront. Among those twelve thousand spectators was Robert W. Gardner, a Scots immigrant, who was working in a machine shop on the northwest corner of Sixth and Maine, just blocks from Washington Square.

Gardner and the crowd were treated to one of the more animated of the debates. A collapsing grandstand delayed the start until 2:30 p.m., but the opponents gave the crowd everything they wanted, and more. The gangly Lincoln was very much the aggressor, slamming Douglas for Congress's refusal to fully address the slavery issue. Douglas parried with a defense of his Senate record, charging that Lincoln sang a different tune about slavery in northern Illinois than he did in the state's southern, proslavery counties.

In the end, perhaps no minds were changed. The two candidates were feted by torchlight parades held by their Republican, Democrat, and Whig supporters and departed for Alton the next morning aboard the *City of Louisville*.

Meanwhile, Robert Gardner returned that Thursday morning to the cramped machine shop at the rear of Sixth and Maine. In his office was a drawing of a steam engine governor he hoped would revolutionize American industry.

Robert W. Gardner

Quincy in 1859

The Illinois riverfront community that Robert Gardner arrived in back in 1852 was a thriving commercial hub that tapped the growing Upper Mississippi Valley. Founded in 1822 by John Wood, later the governor of Illinois, Quincy grew slowly during the 1820s and 1830s, surviving the Blackhawk War and an outbreak of cholera in successive years in the early 1830s.

During its early years, Quincy was host to a number of residents who would make their mark in Illinois politics. Wood, the community's founder, was mayor of Quincy several times before serving as governor of Illinois in 1860 and 1861. Orville H. Browning, a longtime friend of Abraham Lincoln and an attorney in Quincy during the 1830s, 1840s, and 1850s, served as U.S. senator from Illinois from 1861 to 1863. Browning's successor as U.S. senator, William A. Richardson, was from nearby Rushville. The two collaborated in the mid-1840s in defending the ringleaders of the mob that killed Mormon leader Joseph Smith in nearby Carthage.

Stephen A. Douglas, Lincoln's opponent in the 1858 debates, practiced law in Jacksonville, eighty miles east of Quincy. Douglas's death in 1861 opened the Senate seat for Quincy's Browning.

Quincy's downtown was substantially smaller in 1858 than it is today, but befitting a river town, the city grew up the hill from the municipal docks and freight wharves on the riverfront. The Quincy House, opened in 1838 at Fourth and Maine, catered to the rail and steamboat passengers who flowed through the community in a never-ending stream. The rest of Maine Street downtown was fronted by a group of impressive-looking brick and stone business buildings, some as much as four stories tall, housing everything from hardware stores to an organ company.

Steamboats plied the Mississippi River and were the major transportation venue for passengers and goods. *Courtesy of the Historical Society of Quincy and Adams County*

Civic boosters in 1858 took great pride in regular service offered to the Gem City by the Great Western Railroad, and the city was a regular stop for the steamboats plying the Upper Mississippi River. Most hauled freight, but some of the steamboats—called packet steamers—hauled both freight and passengers. Protected by a levee and natural harbor, Quincy had daily packet service to St. Louis by the mid-1850s, as well as packet service north to Galena and St. Paul.

Steamboat freight service from Quincy primarily hauled agricultural commodities and livestock to the markets in St. Louis and beyond, even to New Orleans. Typical shipments were pork, flour, and wheat. Later, the steamboat would figure heavily in the fortunes of the Gardner Governor Company.

John Robertson, Robert Gardner's partner from 1859 to 1870

With Germans making up nearly one-third of the population of Quincy, breweries were a thriving industry by the 1850s. Because of its isolation, Quincy in the 1850s did boast a major carriage and wagon manufacturing business to supply customers in Quincy, Hannibal, Missouri, and the immediate area. One of the biggest industries in the Gem City was the manufacture of stoves, used both for cooking and for heating residences and businesses. The first iron foundry in the community was established in 1846, and by the mid-1850s, the foundry's stove works was manufacturing nine thousand stoves a year and employing sixty people.

What would become Quincy's largest industry was founded in 1859 when Robert W. Gardner set out on his own to manufacture a simple but indispensable component for America's industrial revolution.

The Gardner Governor

The steamboats that plied the Mississippi River above and below Quincy were one of the more visible manifestations of the industrial revolution that had crossed the Atlantic Ocean from Great Britain. Steam engines drove the rapid industrialization of America and Europe in the first half of the nineteenth century. They typically ran on abundant fuel, and as one anonymous observer pointed out, "could do the work of a hundred men."

James Watt, the Scotsman who had improved on Thomas Newcomen's first steam engine in the 1770s, had developed a governor that helped keep the engine from running wild, seizing up, or even exploding. The Watt governor was a crude design with a gravity-type head, usually mounted on the engine frame. The head was then connected by a rod to a butterfly valve often located several feet or more from the gravity head mount.

When the steam engine came to the new United States late in the eighteenth century, American tinkerers immediately began experimenting with improvements to the governor system. Early American steam engines used a dual-disc type valve placed in a chamber directly connected with the governor head. In 1848, an inventor in Rochester, New York, used a gravity-type ball head with a ball of sufficient size and weight to pull the governor valve wide open and thus control the speed of the engine.

Eleven years later, Robert Gardner of Quincy designed a gravity-type flyball governor that improved on the performance of the governor invented in New York. With machinist John Robertson, Gardner established the Union Machine Shop and Foundry on South Fifth Street. In 1859, Gardner spent most of the year on the design of the new governor. On August 14, 1860, he received his first patent from the U.S. government for what he called the Gardner Flyball Governor. The patent explained the principle of using the centrifugal force of the balls to close the valve and the gravitation of an adjustable weighted lever to open the valve. In essence, the patent noted, the two forces worked against and independent of each other. During the remainder of the 1860s and 1870s, Gardner applied for and was awarded additional patents that gave impetus to the growing business, and the manufacture of governors became a rapidly growing business.

Early Years

In the mid-1860s, the firm, which was then known as Gardner and Robertson, manufactured Gardner's Governor and Automatic Safety Stop. To diversify the business, Gardner and Robertson opened a machinists and manufacturers supply store in the 300 block of Maine Street in 1867. Two years later, the partners purchased the Lorenzo and Charles Bull Hardware Store in the 400 block of Maine Street.

In 1870, the renamed Gardner Steam Engine Governor Shop and Office was located on the northwest corner of Fifth and Ohio streets. That year, Gardner bought out Robertson, his partner, who soon after moved to California. After that point in time, Gardner concentrated on manufacturing steam engine governors. The last quarter of the nineteenth century saw the increasing development of steam-driven equipment, and one of the great

Early flyball governor used on steam engines

Gardner Governor Works on the southeast corner of Fifth and Ohio

Advertisement promoting Gardner's first commercial shop

Hardware store located at 424 Maine Street

contributions to the utilization of steam was the perfection of the steam engine governor. The Quincy firm sold more than 150,000 governors to customers across the United States and Canada between 1859 and 1900.

The early 1880s were a critical time for the growing Quincy manufacturer. In 1880, the Gardner plant and its thirty employees were manufacturing 1,200 governors a year. But on July 28, 1880, the Gardner plant was destroyed in a massive fire. Robert W. Gardner vowed not to give up in the face of adversity. The day following the fire, he engaged the services of local architect William B. Hauworth, who quickly drew up plans for a new, expanded Gardner Governor plant on the southeast corner of Fifth and Ohio streets.

In 1883, the firm, which was known to most Quincy residents as Gardner & Robertson, was incorporated as the Gardner Governor Company, the name under which the company operated for the next forty years.

That same year, J. W. Gardner began to become a real factor in the operation of the new Gardner Governor Company. The son of Robert Gardner, he helped secure

Employees pose outside of the Gardner Governor Works.

J. W. Gardner, son of founder, Robert W. Gardner

a basic patent on a new type of governor known as the spring governor in 1883. As superintendent of the shop, young J. W. Gardner assumed his first managerial responsibilities in the company that he was destined to see develop into a large manufacturing plant. Among his most valuable keepsakes was a medal awarded to the Gardner Governor Company in 1883 at the Southern Exposition in Louisville, Kentucky, for the best steam engine governor built in this country.

A New Plant

The company's early successes led to the construction of a new plant on the Quincy waterfront. The location, south of downtown along South Front Street, was in an area that the editor of the *Quincy Daily Journal* called "a sickly swamp and low land." The newspaper's editor, Hiram Wheeler, was championing a new industrial park northeast of the city, and he told his readers that the Gardner Governor site was so waterlogged that passenger trains crawled through the area

THE QUINCY D

VOLUME XXXVII--NO. 243. QUINCY, ILLINOIS, THURSDAY

M ANUFACTURERS AND

S. NICHOLS. J. W. SINNOCK.

NICHOLS & SINNOCK,

Importers and Wholesale Dealers in

Notions, Furnishing Goods, Etc.,

Nos. 213 and 215 N. Third.

Quincy. - - - Illinois.

ISAAC LESEM & CO.,

WHOLESALE

DRY GOODS

—AND—

NOTIONS.

Stock Large and Complete. Prices Always the Lowest.

J. B. Schott Saddlery Co.,

...rers and Dealers in

...HER,

...ardware,

...s,

...h and Mohair ...ep the largest ...ve goods. We

AUSTIN & KOHL,

Wholesale Grocers, Agents For Mott's Celebrated

N. Y. State Cider

—AND—

American Powder Co's

"DEAD SHOT" POWDER.

317 & 319 Hampshire St. : : Quincy, Ill.

ESTABLISHED 1858.

S. E. SEGER'S SONS.

WHOLESALE

GROCERS,

QUINCY, ILL.

Capital Stock $100,000, Paid Up.

Warfield Grocer

COMPANY.

34-36 N. 3rd St. Quincy, Ill.

The first exclusive Wholesale Grocery House in Quincy. Jobbers of all grades Cigars Agents for Stratton & Storm's goods.

NO GOODS SOLD TO CONSUMERS.

Agents for Miami Powder Company.

W. S. WARFIELD, Pres, J. A. PARKINS, V. Pres. C. A. BLANK, Sec, and Trea.

H. A. WILLIAMSON,

COMMISSION ...ERCHANT,

THE USE OF "ECONOMY" STOVES

BRINGS COMFORT & PLENTY

—MANUFACTURED BY—

Comstock, Castle & Co.

Who make three hundred other kinds and styles of Cook and Heating Stoves.

ESTABLISHED 1849.

QUINCY, : : : : : : ILLINOIS.

THE F. W. MENKE

Stone and Lime Comp'y,

—CONTRACTORS FOR—

Cut and Machine Sawed

Lime and Sandstone.

GARDNER PRINC & GRAVITY

Governors.

Over 50,000 in use all over the world the best evidence of their value.

Used exclusively by nearly all the leading engine builders of the United States.

The leading Governor and the largest establishment of the kind in the world. Every Governor guaranteed to give perfect satisfaction or money refunded. Illustrated catalogue and price list. Address

... GARDNER GOVERNOR CO.,
QUINCY, ILL., U. S. A.

...LLER & CO

This ad appeared in 1895 in the *Quincy Democrat* newspaper.

Front Street construction in 1901

"crotch deep" in water, with the locomotive's firebox propped up on a wood frame to keep it from going out. Robert W. Gardner himself had called the site a "beautiful swamp," but he needed a spot close to the riverfront where he could expand his growing business. By 1900, the plant at Fifth and Ohio and its 160 employees was proving totally inadequate to the task of meeting orders. Gardner Governor estimated that the company's business could easily justify employing up to four hundred workers.

Accordingly, the Gardners purchased five acres of ground on the bottoms south of Quincy near the Williamson Oil Plant and the old Curtis Distillery. Gardner Governor announced plans to build a three-building complex on the site, including a main building three hundred feet in length. The buildings were constructed of steel and stone and included three railroad spurs alongside each building as well as through the lengthy machine shop.

General Contractor Buerkin and Kaempen turned over the $60,000 complex early in the spring of 1901. The Gardners threw a memorable grand opening party on April 15, 1901, for more than one thousand invitees. The floor of the machine shop was waxed for dancing to the music of Louis Rischar and his fifteen-piece Empire Theatre Orchestra. Both Gardners were on hand to greet guests, as well as superintendent Robert A. Kiefer.

The Gardner Governor Company weathered a major labor upheaval in 1903 and continued to grow its business through the first decade of the twentieth century. The company celebrated its fiftieth anniversary in 1909, two years after the founder, Robert W. Gardner, died in Quincy. By 1909, Gardner products were in great demand in American industry.

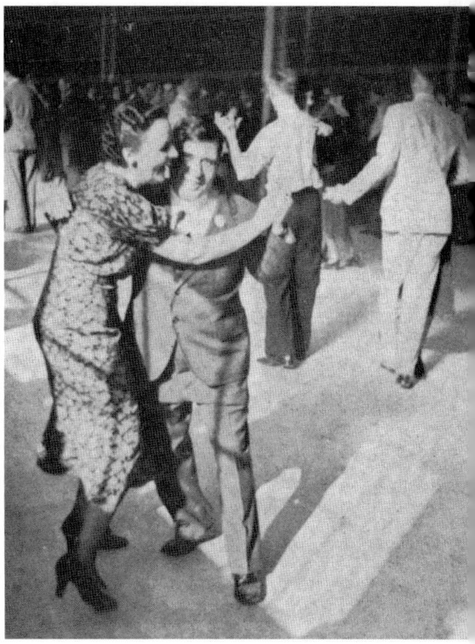

The completion of each new building warranted a dance at the plant.

Robert Gardner poses with workers along Williamson Street at the Front Street plant.

A couple of significant events happened at the turn of the century that would impact the future of Gardner Denver. One was the experimentation of "horseless carriages" driven by gasoline or kerosene reciprocating piston engines. When cars were introduced into the mainstream of American life, the demand for oil increased. The second event was the discovery of oil at Spindletop, a gusher on the Gulf Coast south of Beaumont, Texas, which ushered in America's first oil boom. With the demand for oil high, equipment would be needed to help bring it up from the ground. The Gardner drilling pumps, or mud pumps, were developed for that purpose.

What helped the Gardner Governor Company was the fact that the Mississippi River flowed right by the company's doors. Crates of mud pumps could be loaded onto barges, floated downriver to New Orleans, and off-loaded onto coastal packets for the short trip over to Beaumont. Drilling pumps are manufactured to this day, and the company's petroleum equipment division remains a mainstay of Gardner Denver Inc.

Spindletop oil gusher. *American Petroleum Institute Photograph and Film Collection, Archives Center, National Museum of American History, Behring Center, Smithsonian Institute*

The mass production of cars created an increased demand for fuel, and gas stations popped up all over Quincy. Pictured here is Richardson's service station. *Courtesy of the Historical Society of Quincy and Adams County*

Artist's rendition of the Gardner Governor Company plant in the early 1900s

The South Fourth Street Line

Streetcars helped bring employees to the Gardner Denver plant. *Courtesy of the Historical Society of Quincy and Adams County*

When Robert W. Gardner announced early in 1901 that his Gardner Governor Company intended to relocate from the firm's plant at Fifth and Ohio Street to a much larger greenfield complex on South Front Street, it was one of the largest economic development events in Quincy's history. Gardner's plans called for doubling the workforce to nearly four hundred people.

But the move nearly didn't happen. Most of the plant's workers lived in residential neighborhoods north and east of the Front Street plant. Accordingly, Gardner asked the Quincy Horse Railway & Carrying Company to extend its streetcar line down South Fourth Street to Front Street and along Front Street past the new Gardner Governor plant.

First chartered in 1865, the streetcar company flatly refused Gardner's request. Quincy Horse Railway had operated mule-powered streetcars for a quarter-century before switching to electric traction in 1891. In 1898, the local investors sold to Western Railways & Light Company, a utility holding company with offices in Portland, Maine.

Gardner took his complaint about the streetcar company's refusal to serve the new plant to Quincy Mayor John Steinbach. If he couldn't get streetcar service to the new plant, Gardner told the mayor, he might have to consider moving Gardner Governor to a community with a better transportation system. The mayor immediately contacted executives of the streetcar company, and he likely reminded them of the city's power to award franchises to public utilities in Quincy.

Soon after, Quincy Horse Railway & Carrying Company announced it would lay track down South Fourth Street and along Front Street past the new Gardner Governor plant. The company agreed to have a car at the plant at 6:30 every morning, followed by cars with trailers at 6:45 a.m. and 6:53 a.m., with service every twenty-four minutes thereafter.

J. W. Gardner took over as president of Gardner Denver when his father died in 1907. Besides financial aid to hospitals and charitable institutions, he gave generously to many other civic organizations, serving as president of the board of Blessing Hospital and as a trustee of Woodland Home. *Courtesy of the Historical Society of Quincy and Adams County*

When Robert Gardner died in 1907, the Gardners donated money to the park district to buy land for a park that, to this day, carries his name. *Courtesy of the Historical Society of Quincy and Adams County*

CHAPTER 2
The Denver Rock Drill Company

For the company that would become Gardner Denver Inc., the year 1859 held a momentous significance. It was in that year that Robert W. Gardner founded a machine shop in Quincy, Illinois, that would dominate the steam engine governor market for the remainder of the nineteenth century. In far off northwestern Pennsylvania that year, Colonel Edwin Drake brought in the first oil well in North America. In later years, the development of the oil industry would provide the basis for marketing and selling the Gardner Governor Company's line of pumps. In Colorado Territory in 1859, prospectors first discovered gold in the high valleys of the Rocky Mountains.

The development of Colorado's mining industry in the latter half of the nineteenth and early twentieth centuries provided a ready market for Gardner Governor's products.

By 1920, the Denver Rock Drill Company was one of Gardner Governor's most productive agents and sold many compressors, which were utilized in connection with the drills manufactured and marketed. The early contact with the Denver Rock Drill Company eventually led to the suggestion that the two companies merge.

The rock drills and sharpeners manufactured by the Denver Rock Drill Company required air for their operation, and the Denver Rock Drill Company offered many possibilities for the increased sales of the products of the Gardner Governor Company. These factors and many others finally resulted in the serious consideration and final culmination of the merger between the Gardner Governor Company and the Denver Rock Drill Company to form what would become the Gardner-Denver Company.

The Evolution of Rock Drills

The Denver Rock Drill Company got its start at the turn of the century manufacturing and selling the Waugh drill, one of the most sought-after pieces of equipment in the Western hard rock mining industry. By 1900, Colorado was one of the world's great mining locations. Since the discovery of gold in 1859, the state had become a leader in the production of gold, silver, and other precious metals. The Cripple Creek Mining District west of Pike's Peak was one of the nation's top gold producers at the turn of the century, and the San Juan Mining District south of Cripple Creek and the Leadville Mining District, located just southwest of Denver, were among the nation's top producers of silver and zinc. The Telluride District and its legendary Smuggler Union Mine in southwestern Colorado was also among the state's most productive gold districts in 1900.

Denver Rock Drill merged with the Gardner Governor Company in 1927.

The Cripple Creek District alone produced $20 million worth of gold in 1900, and by the first decade of the century, Colorado had overtaken California as the nation's top gold producer. Besides gold and silver, Colorado was a major producer of molybdenum, radium, vanadium, and coal. The vast majority of the state's mineral resources were extracted from underground mines that went deep into the earth. The standard method of extraction was to drill a hole into a rock face underground, place explosives in the drill hole, and then crush the rock loosened by the resulting blast.

Above: Logo for the newly founded company

The term "rock drill" was the general designation of a machine to drill holes in rock, but the early designation of a rock drill at the turn of the twentieth century was a drill of the reciprocating piston type, as compared to the hammer type of drill that was later developed. But regardless of the type of drill, a rock drill was typically powered by compressed air.

Early drills were bulky and heavy and often required as many as two or three men to set up and operate. But the two- and three-man drills gave miners a sense of security, especially those who worked in distant drifts of deep miles like those of the Cripple Creek District. As late as 1913, the replacement of two-man drills with lighter, one-man drills in the deep shaft copper mines of Michigan's Upper Peninsula led to a six-month-long strike in the copper district.

But light one-man drills made mines more efficient, and inventors in Denver tinkered with developing and manufacturing a one-man drill that would stand up to the harsh conditions underground.

The Waugh Drill

Fred Iler, the founder of the Iler Rock Drill Manufacturing Company, pioneered the namesake Iler drill in 1905. Lightweight, the Iler drill weighed just over twenty pounds. It was designed as a hand machine and worked on the principle of a carpenter's brace and bit. The machine, however, although only weighing twenty pounds, was too heavy for a man to support and drill holes above the horizontal, and it was not suited to drill down-angle holes, as there was no provision for ejecting the cuttings.

The Iler drills proved to be ultimately unsuccessful, although the firm did manage to sell about six of the drills to mines in the Cripple Creek District. In 1906, D. S. Waugh entered the employ of the Iler Rock Drill Manufacturing Company and designed a drill that was called the "Waugh."

WAUGH DRILLS

FOR

STOPING, DRIFTING
AND SINKING

THESE DRILLS

EMBRACE IN THEIR CONSTRUCTION
SOME ENTIRELY NEW FEATURES,
MAKING THEM SUITABLE FOR ALL
CLASSES OF ROCK BREAKING

COVERED BY PATENTS IN THE UNITED STATES AND
FOREIGN COUNTRIES.

SOLE MANUFACTURERS

THE DENVER ROCK DRILL AND MACHINERY COMPANY

1415-19 EIGHTEENTH STREET
AND 1808-10 BLAKE STREET

DENVER, COLORADO, U. S. A.

409 MONTGOMERY ST.
SAN FRANCISCO.

CABLE ADDRESS: "ROCKDRILL, DENVER"
Western Union or Postal Code, also Special Code in Catalogue

Dan Waugh was an English emigrant from Cornwall, the legendary mining region of Great Britain. His Waugh drill utilized a novel air feed cylinder and piston, which supported and fed the drill itself as the bit penetrated the rock. The Waugh drill greatly speeded the stoping process by allowing drilling of holes upward in the ore body. As a result, mines were able to eliminate manual labor and reduce the tonnage costs of producing ore. The Iler Rock Drill Manufacturing Company soon went out of business, and the Denver Rock Drill Manufacturing Company became its successor.

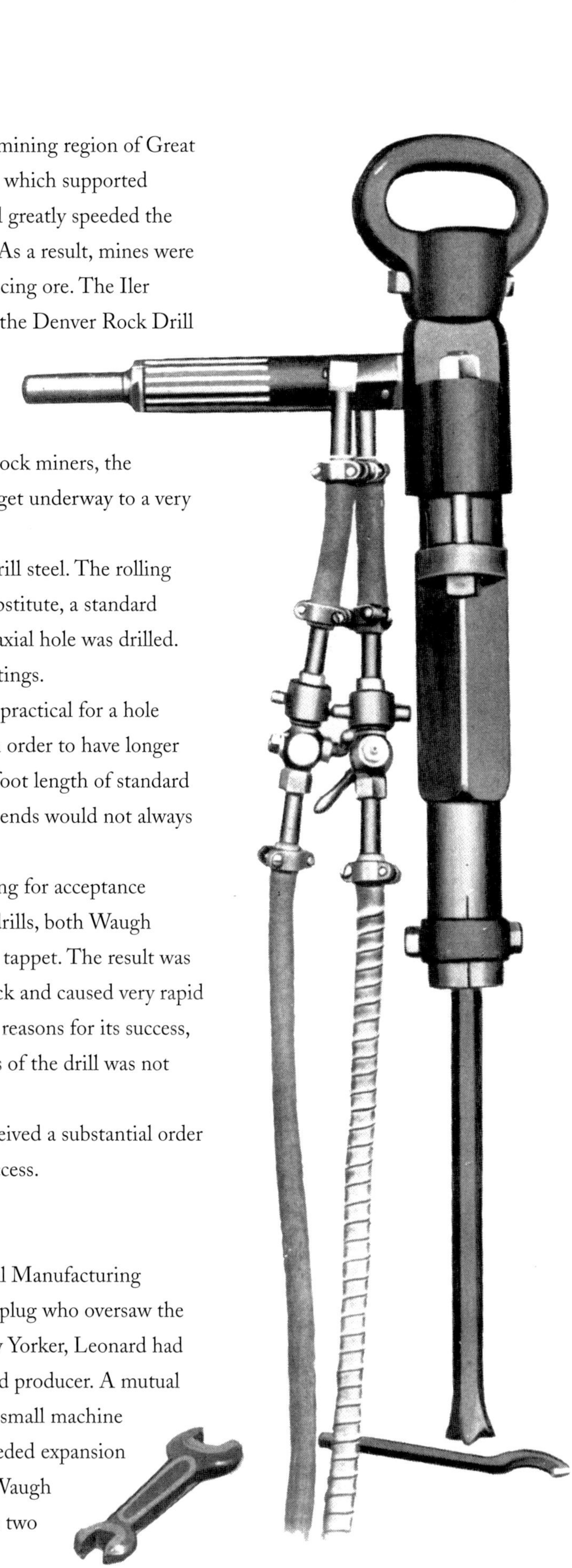

The Waugh drill was a success from its first introduction, and in 1907 and 1908, Waugh designed the Waugh 8C drill. A well-balanced machine that was immediately accepted by Colorado hard rock miners, the Waugh 8C helped the Denver Rock Drill Manufacturing Company get underway to a very successful start.

Denver resident J. George Leyner pioneered the use of hollow drill steel. The rolling of hollow drill steel as used today was unknown at that time. As a substitute, a standard section of octagon or hexagon drill steel was used through which an axial hole was drilled. Through an axial hole, he commingled air and water to expel the cuttings.

A rifle-boring machine was used to drill the hole, but it was not practical for a hole exceeding three feet in length. Company engineers discovered that in order to have longer steel available, they had to drill the hole from both ends, up to a six-foot length of standard drill steel. This was a hit-or-miss job, and the hole drilled from both ends would not always meet squarely. The offset frequently induced breakage at that point.

The Waugh 8C drill was one of a number of rock drills competing for acceptance in the mines of the Rocky Mountain West. The Murphy and Shaw drills, both Waugh competitors, were valveless type drills, but they did not incorporate a tappet. The result was that the cutting from an up-angle hole got into the cylinder and chuck and caused very rapid wear. The introduction of a tappet in the Waugh drill was one of the reasons for its success, although the question of cuttings getting back into the working parts of the drill was not entirely eliminated.

Still, when the Denver Rock Drill Manufacturing Company received a substantial order for Waugh drills, the company thought it was well on the road to success.

Harry Leonard Takes Over

Dan Waugh was the inventive genius behind Denver Rock Drill Manufacturing Company, but W. H. "Harry" Leonard was the entrepreneurial sparkplug who oversaw the company's tremendous growth in the early twentieth century. A New Yorker, Leonard had made a $50,000 fortune by brokering the sale of a Cripple Creek gold producer. A mutual friend put Leonard in touch with Waugh, who was then operating a small machine shop on Wynkoop Street just south of Fourteenth Street. Waugh needed expansion capital and management expertise to help manufacture and sell the Waugh drill. Leonard agreed to invest in the Cornishman's business, and the two became partners in Denver Rock Drill Manufacturing Company.

The company had leased the ground floor of a building at 1415–1419 Eighteenth Street. The tool equipment consisted of an engine lathe, drill press, a power hacksaw, and a grinding attachment for the lathe. Later, Denver Rock Drill Manufacturing Company bought two or three more machine tools, but the shop had no gauges, relying on micrometers and calipers for finished dimensions. Once it went into full production of the Waugh drill, space was soon at a premium.

"We originally only leased the ground floor of the building on Eighteenth Street," Leonard recalled, "and then we leased the second floor. When that did not give us sufficient additional space, we leased a building across the alley facing on Blake Street, then occupied by the Bogue Lead Company. In order to connect the Eighteenth Street shop, we built a passageway over the alley to our additional new quarters facing on Blake Street. We then thought that we had all the space that we would ever require."

The expansion on Eighteenth Street was really only good for about three years. By 1909, Denver Rock Drill Manufacturing Company was seeking a more suburban location in which it could spread out its sprawling manufacturing facilities. That year, the company bought thirty-five lots between Williams and Gilpin streets, and Thirty-ninth and Fortieth avenues. Construction soon began on a shop and office building, which would serve the company for the next half-century. Leonard was particularly proud of the two-car garage, which housed automobiles for the executive and sales departments, at a time when most Denver streets were still dirt. In 1912, the company built a small forge shop on the site to house an Erie Steam Hammer.

The International Steam Pump and Machinery Company, Ltd.

Leonard proved himself more than capable of dealing with the high-powered capitalists and financiers who dominated the North American mining industry in the early twentieth century. In 1912, the Guggenheim Brothers created the International Steam Pump and Machinery Company, Ltd., one of the first modern conglomerates. Daniel S. Guggenheim operated the family's American Smelting & Refining Company (ASARCo), which controlled much of the nation's minerals processing capacity. His brother, Simon, was a United States senator from Colorado from 1907 to 1913.

The creation of International Steam Pump and Machinery Company, Ltd., was an attempt to control companies that sold equipment to ASARCo, as well as to allow the parent company to make a profit on machinery and equipment sold to other mining firms. The Denver Rock Drill Manufacturing Company soon came to the attention of the Guggenheims.

International Steam Pump and Machinery made an attractive offer to Leonard, and he agreed to stay on with the conglomerate as the general manager of Denver Rock Drill & Machinery Company, Ltd., a subsidiary of the parent firm. Leonard might well have taken his profits from the sale of the company and gone on to other endeavors save for unforeseen circumstances. In 1913, shortly after the establishment of International Steam Pump and Machinery, the global economy began a yearlong contraction as bankers became more and more concerned that the Great Powers in Europe were preparing for war. Then, as now, credit dried up, and even the Guggenheims found it difficult to finance their new conglomerate, which depended on cheap debt for its growth.

Leonard saw an opportunity to re-purchase the company he had just sold to the Guggenheims. When he learned that the new owners had failed to record property transfers and mortgages for the purchase, he approached the Guggenheim interests to ascertain their interest in selling. Because of the deteriorating economic conditions, the Guggenheims were more than willing to sell. Leonard bought the assets of the company and reincorporated it as the Denver Rock Drill Manufacturing Company.

The Dreadnaught and the Clipper

With the reincorporation and the move to the new location, Denver Rock Drill began broadening its product line. The company introduced the 16V Valveless Stoper, which quickly began to supplant the Waugh 8C in Rocky Mountain mines. Denver Rock Drill followed up the introduction of the 16V with two drifter drills known as the Dreadnaught and the Clipper. "These were also valveless machines," Leonard explained, "and were mounted on the conventional guide shell. Both of these drifter drills could be used as hand-sinkers, but they were heavy and had poor riding qualities, so that as hand-held machines they were not universally popular."

But the two new machines quickly found employ on boring jobs. "The No. 60 Dreadnaught was very generally used as a Drifter," Leonard said, "and one of the biggest jobs that it made a record on was the Rogers Pass Tunnel for the Canadian-Pacific Railroad in Canada. The contractors on this tunnel made a time bonus of approximately one million dollars, and the efficiency of the Dreadnaught drill was generally accorded as a contributing factor."

When contractors working a mine site east of Eureka, Utah, established a record in 1921 for sinking the Water Lily Shaft more than 425 feet deep in thirty days, they used a Denver Rock Drill Clipper for most of the work.

Compressed air powered the drills.

The Roaring Twenties

Because of the growing demand for strategic minerals during World War I, the mining industry in the Rocky Mountain West and Denver Rock Drill Manufacturing Company continued to grow and prosper. In 1920, Leonard bought additional property adjacent to the company's original holdings. "This represented a full block between Gilpin and Franklin streets, as well as Thirty-ninth and Fortieth avenues," Leonard explained, "and we received permission from the city administration to vacate the streets and alleys running from Thirty-ninth Avenue to Fortieth Avenue. This gave us a solid block of ground with no streets or alleys."

The company was one of the first manufacturers to purchase a set of Johansson gauges. At the time there were only a few manufacturers in the United States that boasted Johansson gauges. Made in Sweden by C. E. Johansson AB, the Johansson gauges were used for calibrating snap gauges and other precision gauging tools. The company boasted that the surface of the Johansson gauges was so accurate that when placed one against the other a molecular attraction was created, holding them securely one next to the other.

By 1926, Leonard had been running Denver Rock Drill Manufacturing Company for almost twenty years. The company had begun efforts to manufacture a portable air compressor on skids to accompany its drills and stopers, but had enjoyed little success. When J. W. Gardner, the president of Gardner Governor, approached Leonard about a possible consolidation, he was quick to see the possibilities.

The two firms merged as the Gardner-Denver Company in early 1927. When Gardner stepped upstairs to the chairman's office, Leonard was named the president of the merged company. With the beginnings of a worldwide depression little more than two years away, Leonard and Gardner-Denver had their work cut out for them.

One of the earlier flyball governors with the new GD entwined logo cast into it.

Many of Gardner's early compressors were used in garages to provide air to fill tires.

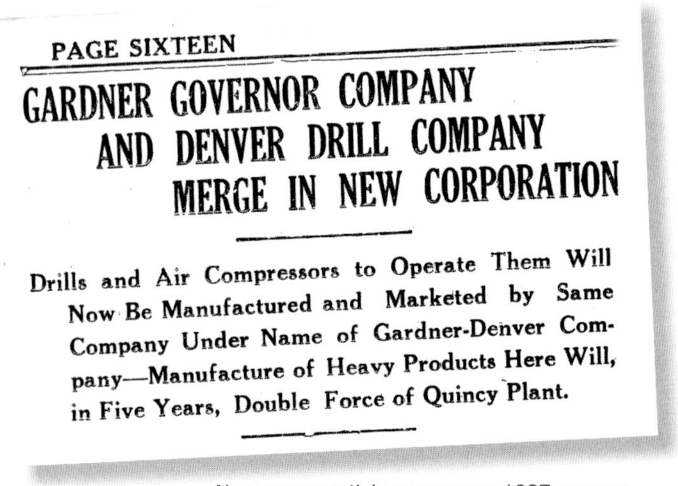

PAGE SIXTEEN

GARDNER GOVERNOR COMPANY AND DENVER DRILL COMPANY MERGE IN NEW CORPORATION

Drills and Air Compressors to Operate Them Will Now Be Manufactured and Marketed by Same Company Under Name of Gardner-Denver Company—Manufacture of Heavy Products Here Will, in Five Years, Double Force of Quincy Plant.

Newspaper article announces 1927 merger.

The Irrepressible Mr. Leonard

The Rocky Mountain West in American history and culture is replete with legendary characters. Wild Bill Hickok. Annie Oakley. Billy the Kid. Butch Cassidy and the Sundance Kid.

W. H. "Harry" Leonard, the founder of the Denver Rock Drill Company and the third president of Gardner-Denver Company, was every bit the Western character himself. Although never a gunslinger, Leonard was representative of the spirit of entrepreneurial capitalism that characterized the mining camps of the region in the late nineteenth and early twentieth centuries.

Born and raised in Mount Kisco, New York, a prosperous rural area north of New York City in the Hudson Valley, Leonard headed west to seek his fortune shortly after dropping out of Harvard Prep School in Groton, Connecticut. He arrived in Colorado's rough-and-tumble Cripple Creek Gold Mining District and quickly secured work with the Cripple Creek Stage Line. Leonard had no experience, but having grown up with horses on the family estate back in New York, he quickly learned how to handle the reins of a four-horse coach.

Leonard drove the stage from Cripple Creek to the head of the Colorado Midland Railroad at Florissant, along sixteen miles of narrow mountain road. He came to know many of the mining magnates, bankers, and railroad executives who controlled the production of gold in the Cripple Creek District, including Albert E. Carlton, Spencer Penrose, and Charles MacNeill.

On one of his trips, Leonard learned that the British owners of the Prince Albert Mine, about halfway between Cripple Creek and Victor, wanted to sell their interest in the property. Leonard approached Carlton, who controlled the First National Bank of Cripple Creek, whom he knew was in the market to purchase a mine. Leonard offered to broker the sale of the Prince Albert, and his commission on the $500,000 purchase came to $50,000.

When Bill Haywood and the Western Federation of Miners went on strike against the Cripple Creek mines for the second time in ten years in 1903, the district was consumed by violence and anarchy. Leonard left for Denver, never to return.

In the Mile-High City, an old friend introduced Leonard, then just twenty-one years of age, to Daniel Waugh. The two men quickly hit it off, and Leonard invested part of his $50,000 grubstake in helping found the Denver Rock Drill Manufacturing Company. By the World War I era, Waugh drills, drifters, and stopers were a fixture in mining districts all over the Rocky Mountain West, not to mention some of the major mining regions overseas.

Leonard continued to provide leadership to Denver Rock Drill Manufacturing Company, and when the Colorado firm merged with the Gardner Governor Company in 1927, he was named first president of the consolidated operations. Leonard retired from Gardner-Denver in 1935 and spent the rest of his long life raising cattle on a ranch near Denver. He died in Denver in 1957 following a stroke.

CHAPTER 3

Gardner-Denver in Depression and War

The merger that created Gardner-Denver Company in 1927 was well planned and well executed. It joined a company that enjoyed a reputation as one of the best-run manufacturers of rock drilling equipment for the world's mining industry with a firm that dominated its niche of manufacturing portable air compressors to the construction industry. The synergies were obvious.

Portable air compressors powered rock drill equipment. The vertical integration of the two manufacturers immediately resulted in increased volume for both divisions of the company because of the potential in the construction industry.

Portable air compressors and rock drill equipment were in growing demand by a new segment of the construction industry in the 1920s. Henry Ford had put America on wheels with the introduction of the Model T, and by the 1920s, state and city governments across the nation were building roads to handle the increased amount of automobile traffic. By 1927, there were fifteen million Model Ts on the road, not to mention millions of other Packards, Studebakers, Oldsmobiles, and the like.

The 1924 Federal Road Act poured millions of dollars of federal money into state highway department budgets. By the end of the 1920s, highway construction was the largest single private industry in North America. The corresponding rise of the oil industry during the decade also proved to be very beneficial to Gardner-Denver's Quincy Division, which saw demand for mud pumps for drilling rigs continue to increase. The passage of the Federal Power Act in 1920 encouraged utilities to embark on major expansions of hydroelectric power facilities.

For Gardner-Denver in 1927, the future was bright. The company reported nearly $300,000 of cash on hand and a bank line of credit totaling $450,000. Customer accounts and notes receivable added up to $965,000, and inventories of finished goods amounted to nearly $2.4 million. Plant and equipment was appraised at $3.45 million. All told, the merged company reported the excess of assets over liabilities totaled more than $7 million in the spring of 1928.

Road under construction in Yosemite National Park

Growth of the Quincy Division

Much of the growth for Gardner Governor during the decade or so preceding the merger had, in fact, come from the manufacture of pumps for the oil industry. Among the pumps manufactured by the Gardner Governor Company in 1913 was an 8x5x10 turret type mud pump. E. F. Schaefer, Gardner-Denver's president during the 1930s and 1940s, recalled that "the first of these pumps utilized in drilling oil wells with rotary equipment were sold by the McNeal Machinery Company of Joplin, Missouri. For a number of years, these 8x5x10 pumps were the best available pumps for this unusual service and were the original 'mud pumps' in the oil fields."

As rotary drilling gradually began to replace cable tool drilling, and more capacity and higher pressures were required in the pumps utilized in the oil fields, the Gardner Governor Company in the early 1920s began to develop pumps for this growing but specialized service. In 1923, the company introduced the 14x6.75x12 slush pump in the California oil fields as the largest slush pump available for oil field service.

The early contacts of the Gardner Governor Company with the oil industry resulted in Gardner's continued interest in serving the oil industry, as it rapidly became a tremendous factor in the growth of the United States. Largely because of its interest in the oil fields, Gardner Governor's Engineering Department in 1923 introduced the first enclosed type eccentric driven duplex power pump. Before the development of the Gardner enclosed type power pump, duplex power pumps were typically of open frame, crankshaft-driven construction. "This type of pump was not satisfactory when exposed to weather conditions," Schaefer explained in the mid-1940s, "and in the oil fields and for railroad service, power pumps had to be installed in outlying places where no shelter was available. The enclosed type eccentric driven pumps introduced by the Gardner Governor Company revolutionized the power pump business, and now an open frame power pump is an antiquated as a Model 'T' Ford. The initial advantage gained in designing this new line of equipment has never been relinquished by the company, and today the Gardner-Denver Company is predominant in the sale of power pumps for slush pump drilling."

Black Thursday

For Gardner-Denver, and thousands of other U.S. businesses, the world turned upside down during the last week of October in 1929. The U.S. economy, which had been fueled since the end of World War I in late 1918 by easy credit and consumer spending, came skidding to a halt on "Black Thursday," October 24, 1929. The beginning of the stock market crash that Thursday, which carried over into the following week's "Black Monday" and "Black Tuesday," wiped out billions of dollars of investment capital and set the stage for the onset of the Great Depression.

Gardner-Denver felt the impact almost immediately. "During the first ten months of the year the volume of business showed a very substantial increase over the corresponding period in 1928," the company wrote shareholders late in the year, "but during the last two months there was a decided falling-off in volume, which no doubt was to some extent a reflection of the stock market condition."

Still, the company saw little cause for concern. The company noted in its 1929 annual report that "keeping our products up to date is most important." The introduction of

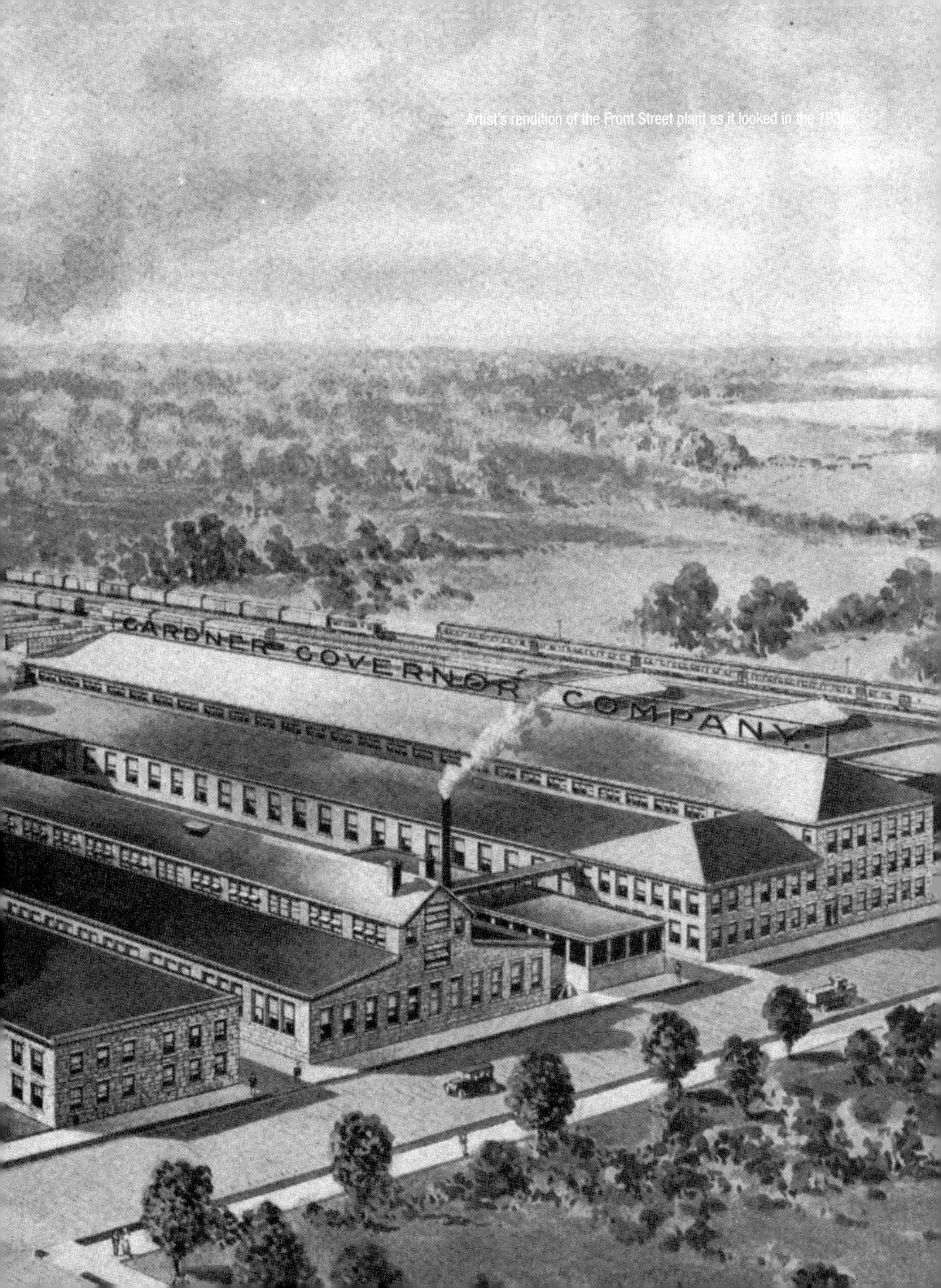

Artist's rendition of the Front Street plant as it looked in the 1930s

power slush pumps weighing up to thirty thousand pounds was an important advance in the provision of equipment to the oil drilling industry, and Gardner-Denver's expenditures for research and development work was nearly $32,500 for 1929, a 20 percent increase over 1928. Business was so good, in fact, that the company spent $16,000 to purchase a tract of land adjoining the Quincy plant and built a $150,000 addition to the office and machine shop.

Above: Aerial shot of the Front Street plant in Quincy in 1937

Right: While streetcars and buses took most employees to work, a few executives drove in their own cars. This picture was taken in the 1930s.

Consolidated net profits of $1.5 million in 1929 dropped to $650,000 in 1930 as the effects of the Great Depression took hold. The company continued to spend money to increase the efficiency of the Quincy plant, installing a number of large forming machines during 1930. The new, more efficient machines resulted in a substantial increase of pump inventory, but the company remained confident that "as business conditions improve, these new products should materially increase our volume of business."

A Worldwide Depression in Business

But at the same time, there were troubling signs that the economy was showing significant weakness. The continuing low price of metals had resulted in mine closings across the American West, resulting in a sharp slowdown of business for the Denver Division. And by the end of the year, demand for oil was calling into question some of the earlier more optimistic predictions about the company's market for mud pumps.

Business continued to slump in 1931 and 1932, although Gardner-Denver did manage to eke out a $150,000 net profit in 1931. Foreign currency devaluation, a collapse of most of the company's mining business, and an overproduction of oil led Gardner-Denver to effect the most stringent economies. Plant closings and layoffs during the year helped the firm reduce total overhead expenses by nearly $950,000. Gardner-Denver, however, pledged to continue investing in plant and equipment in

spite of the worldwide depression in business.

The election of Franklin Delano Roosevelt as U.S. president in fall 1932 brought a measure of optimism to the U.S. economy. Roosevelt, who was not inaugurated until March 1933, immediately shut down the nation's banks to stem a run on the financial system. The "first 100 days" of the New Deal put hundreds of thousands of Americans to work on federal public works projects, and FDR rushed economic legislation to Congress.

For Gardner-Denver, the improved economic conditions meant an immediate uptick in business. The company's net loss of $75,000 through the first five months of 1933 was more than offset by a net profit of $170,000 for the last seven months of the year. The firm completed the year with a net profit of almost $96,000.

A noticeable revival of mining activity and signs that

Construction of Route 66 near Flagstaff, Arizona

the oil business was about to resume drilling followed the next year. Even more important for Gardner-Denver's plans to take advantage of an improvement of business conditions was the firm's current assets to current liabilities ratio, which stood at just over 15:1 when many companies operated with a ratio of 1:1 or lower.

Recovery

Business continued to improve in 1934 and 1935. Net profits jumped to $340,000 in 1934 and more than doubled to $717,000 in 1935. Part of the reason for Gardner-Denver's strong showing in mid-decade was the effect of the massive infrastructure projects initiated by the Roosevelt administration to get the nation back on its feet. Hydroelectric power projects on a scale never before seen on the North American continent helped jumpstart the U.S. economy at mid-decade.

In 1935, in far-off northeastern Montana, thousands of workers swarmed over the Missouri River bottomlands at Fort Peck, building the largest earth-filled dam in the world. Halfway across the continent, more thousands of workers were completing the Wheeler and Norris dams on the Tennessee River in Alabama and east Tennessee, the first of dozens of dams built by the Tennessee Valley Authority. In Black Canyon of the Colorado River on the Nevada-Arizona border, workers were putting the finishing touches on Hoover Dam, perhaps the most iconic infrastructure project of the 1930s era. More thousands of workers were employed on highway, sewer, airport, and municipal building projects underwritten by the Works Progress Administration (WPA), the Roosevelt public works vehicle that was symbolic of the recovery from the Great Depression.

Dams and heavy civil engineering projects required large amounts of compressors, pumps, and drilling equipment. Gardner-Denver stood ready to meet that need, and the massive public works projects of the era would continue to spark Gardner-Denver sales increases into the early 1940s.

A New Man at the Top

After a successful eight years as president of Gardner-Denver, Harry Leonard moved into the more ceremonial position of chairman of the board when J. Willis Gardner retired in 1935. Gardner, who had been associated with the company since joining his father in management in the 1890s, remained active as chairman of the Executive Committee of the Board of Directors.

Leonard, who had expressed the desire to raise cattle on his ranch north of Denver, continued to serve as chairman of the Board of Directors until his retirement in early 1947. Named to succeed Leonard as president was a Quincy Division veteran. H. Guy Myers had joined the Gardner Governor Company in the 1910s, and when the Quincy firm merged with Denver Rock Drill Manufacturing Company in 1927, Myers was named executive vice president of the consolidated company. Leonard, Gardner-Denver's first president, maintained offices in the Mile High City while Myers was responsible for the day-to-day operations of the compressor and pump operations in Quincy. When Leonard moved up to chairman of the board in the spring of 1935, Myers was a natural for the company's top operations job. Myers guided

Construction of the Davis Dam on the Colorado River

Going Public

For most of its first three-quarters-of-a-century of operation, Gardner-Denver and its predecessors were privately held. When Gardner Governor and Denver Rock Drill Manufacturing Company merged in 1927, the new company filed incorporation papers in Delaware and issued 30,000 shares of preferred stock valued at $100 per share to the owners. The company also authorized the issuance of 250,000 shares of common stock at no par value.

The onset of the Great Depression in 1929–1930 dictated the preservation of working capital, and as a result, Gardner-Denver did not pay dividends on either preferred or common stock for a number of years early in the decade. In fact, it wasn't until the summer of 1933 that the company was able to pay a partial cumulative dividend of $4 per share on the preferred stock. During the next year, the company caught up its cumulative preferred payment and began paying a twenty-five-cent dividend on common stock in the summer of 1934.

With business picking up in 1935 and 1936, Gardner-Denver began to investigate the feasibility of going to the equity markets with an initial public offering. In June 1936, the company amended its certificate of incorporation to offer 40,000 shares of $3 cumulative preferred stock at $20 par and 250,000 additional shares of common stock. A. G. Becker underwrote the issue, and on June 20, 1936, 40,000 shares of Gardner-Denver preferred and 35,099 shares of common stock began trading on the Chicago Stock Exchange.

Just under a year later, the board authorized the issuance of an additional one million shares of common stock, and split the existing common 3–1. Gardner-Denver remained listed on the Chicago Stock Exchange until September 28, 1943, when the board authorized the management team to proceed with a listing of the company's common stock on the New York Stock Exchange.

H. Guy Myers became president of Gardner-Denver in 1935.

Gardner-Denver's fortunes through the next twelve years, replacing Leonard as chairman of the board in 1947.

The first part of Myers's tenure was marked by continued growth in the face of still difficult economic conditions. The nation's recovery, which had picked up steam in 1935 and 1936, slammed on the brakes in 1937 and 1938. By adhering to a strict policy of frugality, Myers was able to report record profits of $1.15 million, an increase of almost $100,000 over 1936 profits. "The year 1937 was one of the best in the history of the Gardner-Denver Company," Myers announced in the 1937 annual report, "both in volume of sales and net earnings."

The company spent $1.5 million in 1936 and 1937 on new buildings and plant equipment. By mid-year 1937, Gardner-Denver employed 1,950 people.

The bear market that depressed stocks more than 40 percent and plagued the U.S. economy in 1938 pulled down corporate profits nationwide. Gardner-Denver's net profits in 1938 dropped by nearly half to $590,000, due primarily to a 20 percent decrease in sales revenue for the year. But already, by the latter part of the year, conditions were beginning to improve.

Gardner-Denver's net profits rebounded 90 percent to $1.12 million in 1939 on the strength of sharply better sales and lower unit manufacturing costs. As it had during the mid-1930s, Gardner-Denver used the increased cash flow during the year to make major capital improvements in 1939. The company completed a nearly forty-thousand-square-foot, one-story pattern-making and pattern storage building in Quincy, as well as a substantial remodeling of the existing pattern shop into new quarters for the engineering and drafting departments. The two major expansions at the Quincy site were completed at a cost of almost $270,000.

The company employed 1,300 people, down from the high of 1,950 in 1937 but had hopes of calling back a large number of additional workers if the projected increase in business for 1940 proved to be correct. And in line with its progressive approach to management-labor relations,

Gardner-Denver adopted a plan in 1939 to give two weeks of paid vacation to employees with five years or more of service.

War Clouds

On September 1, 1939, Nazi Germany invaded Poland, and two days later, both France and Great Britain had declared war on the Germans. Imperial Japan and China had been fighting since 1936, and the conflict threatened to engulf more of the war in the new year of 1940.

The United States was officially neutral, but the Roosevelt administration sharply increased spending in 1940 for defense needs. But any increase in sales because of defense contracts was more than offset by higher federal income taxes, which resulted in profits for the year dropping slightly in spite of about a 9 percent increase in sales.

Employment had grown by more than five hundred people since 1933, and the Denver Division was in the process of adding a line of contracting equipment to complement its line of rock drills. In the late 1930s and early 1940s, Denver Division introduced a number of new products, including the G1 Bit Grinder, WD4 Derrick Drill, D89H Drifter, R104 Stoper, D73 Drifter, S73 Sinker, R91SC Stoper, and a line of CF Feed Mountings.

What is now a pump testing facility was constructed in the early 1940s.

As the year 1941 entered its last month, H. Guy Myers and the Gardner-Denver management team looked forward to continued strong growth in business in 1942. On December 7, 1941, Japanese dive-bombers attacked the U.S. Naval Base at Pearl Harbor, Hawaii. The next day, President Franklin D. Roosevelt described the attack as "a date which will live in infamy" and asked the U.S. Congress to declare war on Japan. Before the week was over, Germany and Italy had declared war on the United States.

Gardner-Denver would spend the next forty-four months helping supply U.S. war demands.

Gardner-Denver at War

The surprise attack on Pearl Harbor shocked the country, and Gardner-Denver, like most manufacturing companies throughout the United States, did what it could to help win the war. Some Gardner-Denver equipment was used to help raise many of the ships that had been sunk at Pearl Harbor.

World War II brought about many changes for Gardner-Denver. The company faced tremendous manpower shortages as hundreds of its employees answered the call to duty and served their country on the far-flung war fronts. Employees on the factory floor back in Quincy and Denver coped with shortages of everything from rubber to steel to industrial oils and lubricants. At home, the employees faced shortages of dozens of commodities, including gasoline, meat, sugar, tobacco, and cooking oil.

Employees bought war bonds to help finance the nation's defense efforts. In Quincy, employees participated

New building going up on the east side of Front Street in 1936

World War II production rally at the Quincy plant

in scrap metal drives and helped roll Red Cross bandages for their co-workers overseas. In Denver, employees grew Victory Gardens to supplement the often-sparse supply of foodstuffs available through the rationing system. Employees got together monthly to make up packages of cookies, cakes, snacks, and tobacco for the men and women serving in the armed services. By the end of 1942, 345 Gardner-Denver employees were serving in the armed forces.

More than once during the next three years, the Gardner-Denver family would be saddened to learn of the death of a former employee in combat in the European or Pacific theatres. The World War II era also witnessed the passing of a tangible connection to the founding of the company. On February 1, 1943, J. Willis Gardner, Gardner-Denver's chairman of the Executive Committee and the son of founder Robert W. Gardner, died in Quincy at the age of eighty. Gardner had joined his father in the company in the early 1880s when the then Gardner Governor Company had forty employees. When Gardner died sixty years later, Gardner-Denver's work force was more than two thousand people.

Quincy employees are urged to give their all to the war effort.

Even given the hectic pace of activity during the war, Gardner-Denver managed to build a cut-off shed, a major addition to the shipping building, and a sandblast building, all at the Quincy site. The company also introduced several new additions to its product lines, including the T23 Tamper, the PCO Column, the GD14 Loader, and the B37, B67, and B87 Paving Breakers.

The end of World War II in the spring and summer of 1945 infused Gardner-Denver with confidence about the future. The company had been weaning itself from war contracts since mid-1944 and hoped for a relatively easy time of converting to peacetime production.

Forged in the flames of depression and war, Gardner-Denver looked forward to the opportunities available in a postwar world.

Echoes of a Trip Around the World

J. W. Gardner traveled to exotic places all over the world, sending much-cherished postcards back to employees and vendors.

J. Willis Gardner celebrated his sixtieth birthday in 1923 with a trip around the world.

The primary purpose of the trip was "to investigate business prospects and conditions in the various countries and to further increase the value of our foreign department."

Gardner and his wife Helen left Quincy in the late fall of 1922 and spent most of the next eight months visiting England, France, Egypt, India, Burma, the Malay states, Java, Indochina, China, and Japan. In each of the countries they visited, the Gardners mailed postcards to hundreds of Gardner Governor employees, customers, and vendors.

It was perhaps a simpler world, with English and French colonial empires stretching across much of the globe. Their accommodations from Suez were POSH—Port Outbound, Starboard Home—to minimize the brutal sun of the Red Sea and the Indian Ocean.

They mailed home postcards describing the unloading of coal at the docks in Japan and witnessed the beheading of a criminal in Siam. They visited temples in Indochina and watched fishermen cast their nets into the seas off Burma.

When they returned to Quincy in the summer of 1923, J. Willis and Helen Gardner commissioned a pamphlet describing their travels and mailed it to all of the names on their postcard list. The couple received hundreds of letters from friends, customers, employees, and vendors. These they bound into a three-volume collection entitled *Echoes of a Trip Around the World.*

The trip was so successful that the couple elected to repeat it ten years later for J. Willis Gardner's seventieth birthday. This time, the Gardners started in Japan and China, working their way west against the setting sun. Once again, the Quincy couple kept the Gardner-Denver community apprised of their journeys by postcard, mailing as many as six thousand for each country they visited.

"I suppose this is a sort of advertising for our firm," Gardner explained his habit of mailing postcards to a Filipino newspaperman in 1934, "I enjoy doing this, and my friends enjoy getting them."

CHAPTER 4
A Century of Serving the World's Basic Industries

Gardner-Denver celebrated the return to a peacetime economy with a party. In June 1948, the company celebrated the completion of the new receiving building and the substantial expansion to the company's tool room by hosting an open house on the Quincy campus. On a beautiful early summer Saturday afternoon, more than four thousand residents toured the rambling plant and went through buildings both east and west of Front Street and north and south of Curtis Creek. Many of the residents had never been inside the sprawling factory, which employed many of their neighbors.

Visitors entered the main doors of the offices to start the tour, where they were greeted by a receiving line consisting of President Edgar F. Schaefer; H. Guy Myers, chairman of the executive committee; and a half-dozen other members of the Gardner-Denver executive team. Visitors were guided in their tour by printed maps and by orange cards bearing big black arrows to point the way along the route. At the east end of the vertical compressor assembly department, visitors found long tables laden with cupcakes and ice cold soft drinks, served by more than a dozen attendants.

"Everywhere in the plant, which was as orderly and neat as such a busy big shop should be," wrote a *Herald-Whig*

More than four thousand residents attended the open house in 1948.

reporter, "visitors saw gleaming, machine-cut parts, rough castings, polished fittings that go into the final assembly of the pumps and compressors that find work in mines and oil fields and on heavy construction jobs everywhere."

Helping the World Recover from War

Indeed, Gardner-Denver equipment was in place around the world, as work crews on every continent set about rebuilding an economy that had been ravaged by a decade of war. The company's pumps, compressors, and rock drilling

During times of high pedestrian traffic, a stoplight would be rolled out into the middle of the road. (Notice the electric cord across the road.)

equipment were in huge demand in Western Europe, Australia, and South America.

Trucks used at the plant in Quincy

Increasingly, Gardner-Denver pumps were finding employ in the rapidly developing oil fields of the Middle East. The war had revealed the almost limitless nature of the oil deposits underlying the empty sand dunes of the Kingdom of Saudi Arabia. Aramco, a partnership of California Arabian Standard Oil Company, Texaco, Standard Oil Company of New Jersey, and Socony Vacuum, signed agreements with the Saudi royal family beginning in the 1930s that let the American company prospect for and develop oil fields in most of the eastern half of the Saudi peninsula. In 1950, an Aramco subsidiary, the Trans-Arabian Pipeline Company, completed a pipeline from the Saudi fields to the Mediterranean coast in Lebanon, allowing Saudi oil to be shipped quickly to refineries in Western Europe and the United States.

In the United States, the Gulf Coast oil fields of Texas and Louisiana were in the midst of an exploration and development boom. By the end of 1948, eighty-two development projects along the Texas Gulf Coast had been completed, and another forty-three projects were underway. Meanwhile, wildcatters were ranging across the flat plains of West Texas, mapping the Permian Basin and its boundless deposits of oil and natural gas.

All of the exploration and drilling was being done for a purpose. America's economy in the postwar era was rapidly converting from being fueled by coal to being fueled by oil. Between 1946 and 1950, the demand for petroleum in the United States skyrocketed, with use increasing from 1.75 billion barrels in 1946 to 2.25 billion barrels just four years later.

The continuing growth in petroleum consumption was

An example of Gardner-Denver's service to general industry is shown in this picture of an automobile assembly line where 113 "air feed drills" arranged in a panel drill all the auto body's trim holes in fewer than six seconds.

driven by two major factors. The United States became a petrochemical society in the years following World War II. Plastics became a part of American life in the late 1940s and 1950s, thanks to the almost ubiquitous presence of plastics in consumer packaging and, as the decade wore on, in the building and automotive industries. A second factor driving oil consumption was the tremendous boom in automobile ownership. Household vehicle ownership surged in the postwar years as Americans were finally able to afford cars after a decade of depression and five years of war. Motor vehicle production increased from just over 2 million passenger cars in 1946 to 6.6 million in 1950. Annual production of passenger cars would drop during the Korean War years but would rebound to almost eight million cars in 1955.

The U.S. automotive and petroleum culture was subsidized by federal and state governments that embarked upon a massive program of highway construction in the postwar years. That support was ratcheted up during the presidency of Dwight D. Eisenhower. The National Interstate and Defense Highway Act, coupled with the Federal Aid Highway Act of 1956, created a self-funding mechanism for a federally sponsored system of limited access highways across the United States. The legislation called for the construction of forty-one thousand miles of toll-free, four-lane concrete highways across the nation, paid for by a Highway Trust Fund supported by gasoline, truck, tire, and other automobile taxes. The bill called for lanes at least twelve feet wide and absolutely no intersections.

Work on the
Pennsylvania Turnpike

Road construction near the
Golden Gate Bridge

The dramatic increase in oil consumption was a real boon for Gardner-Denver's petroleum division, which continued to be a major supplier of mud pumps to drilling rigs around the world. But the growth of automobile traffic, and the corresponding push to build new roads and highways, meant even more business for Gardner-Denver, selling rock drills and compressors to the nation's heavy construction industry.

New Lines

In 1949, Gardner-Denver introduced the Twindrill for drilling holes that would then be blasted for laying oil and natural gas pipelines. The Twindrill was quickly adapted for use by Tapline, the Aramco subsidiary building the first major pipeline across Saudi Arabia.

Gardner-Denver Twindrills also were employed in pipeline construction across the Western Hemisphere, on projects from Bolivia north to Alaska.

The company modernized its drill steel sharpening building in 1950 and added the adjacent K Building to facilitate production of the S58 Sinker. Another good seller at the time was the Gardner-Denver Wagon Drill. Drifters, including the DH99 Drifter with sectional rods and couplers, also found favor with contractors building major hydroelectric dam projects in North America, South America, Australia, and Africa.

The growth in business necessitated upgrading the Denver facilities. In 1954 and 1955, Gardner-Denver erected a new building to accommodate the increase of equipment associated with the company's deep hole drilling business. Gardner-Denver also built a new shot blast building, a new oil house building, and new offices for the Denver Sales Branch.

A Change at the Top

The Gardner-Denver family went into mourning on Monday morning, May 3, 1954, when they learned that Edgar F. Schaefer had died suddenly at his home on Maine the previous evening. Then only fifty-eight years old, Schaefer had succeeded H. Guy Myers as president in

A Gardner Denver Twindrill was used for trenching. It helped to cut the cost of laying oil and gas pipelines through rocky formations.

1947 and had guided Gardner-Denver through the rapid expansion of business in the post–World War II years.

"Shaef, as he preferred to be known by all of his friends, had the tremendous capacity for appealing to and holding many friends from all walks of life," said Charles H. Meredith, then the executive vice president of the Industrial Association of Quincy. "His ability to get people together on common problems and to get directly to the point on solving community problems will be sorely missed."

Edgar Schaefer succeeded H. Guy Myers as president in 1947.

A Quincy native, Schaefer was a 1916 graduate of the University of Illinois. He returned to his hometown to teach, but when the United States declared war on Imperial Germany in the spring of 1917, Schaefer joined the U.S. Army. During the war, he served as an Army pilot and saw combat action in France.

Schaefer returned to Quincy following demobilization, and he joined the then Gardner Governor Company as a cost clerk in the summer of 1919. He rose rapidly through the ranks, serving as export manager and assistant sales manager before being named vice president in charge of sales in 1926. The next year, Schaefer was named to the board of directors of the recently merged Gardner-Denver Company.

Gardner-Denver promoted Schaefer executive vice president in 1942, and five years later, in 1947, he succeeded Myers as president of the company. During World War II, Schaefer served as a member of the influential Illinois War Production Board, which established industrial policy for the entire state.

A founding director and first president of the Industrial Association of Quincy, Schaefer's legacy in his hometown was the manufacturing boom that had characterized the community from the beginning of World War II until his death. "The employees of Gardner-Denver Company have lost the understanding and humanness of an outstanding business leader," eulogized Meredith, Schaefer's longtime colleague at the Industrial Association. "The community at large has lost a tremendous personality and a man of great capacity and demonstration of leadership."

Gardner-Denver turned to another sales veteran to replace Schaefer at the helm. Gifford V. Leece had grown up in Bishop, California, southeast of Yosemite National Park and attended the University of Southern California for a year before joining the Army during World War I. Like Schaefer, Leece joined Gardner Governor as a clerk in the years immediately following the end of the war.

Gifford Leece

During the 1930s, Leece was named to head all of Gardner-Denver's West Coast operations, and when the company opened its first export office in New York, Leece moved cross-country to head the new effort. During World War II, he moved back to Quincy and served as vice president and general sales manager, reporting to Edward Schaefer. When Schaefer succeeded Myers as president in 1947, Leece succeeded Schaefer as executive vice president.

The Merger with Keller Tools

As it was, Leece's first big impact as president of Gardner-Denver was domestic. During the first five years of Leece's presidency, Gardner-Denver embarked upon an aggressive expansion program that was driven by mergers and acquisitions rather than internal growth. It was the first merger initiative undertaken by the company since Gardner Governor and Denver Rock Drill Manufacturing Company had begun negotiations nearly thirty years before.

The first successful merger since that between the two namesakes occurred in 1955 when Gardner-Denver purchased Keller Tool Company of Grand Haven, Michigan. A well-regarded pneumatic tool manufacturer, Keller Tool had grown rapidly during World War II with its line of chipping hammers that sold to shipbuilders who were busy constructing the nation's battle and merchant marine fleet.

Leece credited R. G. Gardner, the company's chairman and treasurer, and the grandson of founder Robert W. Gardner, with convincing the Gardner-Denver Board to pursue Keller Tool. Leece recalled that Gardner-Denver had never done business with Keller Tool, but that R. G. Gardner "was the one who thought we should add a pneumatic tool line. He was really the one who inspired the merger."

Keller Tool was a family-owned, privately held business. Leece began negotiations with Vincent "Vin" Erickson, the family member who served as Keller Tool's president. Leece understood the pitfalls that any merger faced.

Fortunately, Leece recalled, Vin Erickson was 100 percent in favor of the merger. It also helped that Gardner-Denver immediately began integrating the talented young engineers and supervisors from Keller Tool into the bigger parent company and its operations. One of those Keller Tool employees who would make a career at Gardner-Denver was Ed Fosheim. Born and raised in the Dutch Calvinist community of Grand Haven, Fosheim earned a BA degree in marketing from Michigan State University in East Lansing in 1951 and then spent three years in the U.S. Army during the Korean War. He joined Keller Tool in 1954, and then completed his MBA at Michigan State in 1957. Fosheim was assigned to Gardner-Denver's Wire Wrap Division and was then asked to relocate to the Quincy offices in 1959.

Fosheim and his wife were among the hundreds of people who were moving into the rapidly growing Quincy area during the mid-1950s. They moved into a house on the east side at a time when Thirtieth Street was the far east end of Quincy and raised their children in the community. Fosheim's career with Gardner-Denver lasted more than thirty years and included twelve years in the corporate offices in Dallas, Texas, during the 1970s and 1980s. He returned to Quincy in 1986 and retired two years later.

Keller Tools in Grand Haven, Michigan

Mayhew and More Mergers

Gardner-Denver followed up the successful merger with Keller Tools in 1955 by the acquisition of the nation's largest producer of seismograph and geophysical drilling equipment for the oil industry. Mayhew Machine of Dallas also manufactured light oil well drilling rigs and rotary drilling equipment for the construction and mining industries. Vertically integrated, Mayhew Machine also owned and operated Mayhew Supply as a sales organization to market the company's products.

Founded in 1923 by H. W. Mayhew Sr., Mayhew in 1958 was run by his sons, H. W. Mayhew Jr. and Charles Mayhew. Gardner-Denver issued 138,000 shares of its common stock in exchange for all of the outstanding stock in the two companies, and H. W. Mayhew Jr. joined the Gardner-Denver board of directors. Mayhew's more than two hundred employees were welcomed into the Gardner-Denver family, bringing employment of the company to nearly thirty-seven hundred as Gardner-Denver headed into its centennial year.

The addition of Mayhew greatly broadened Gardner-Denver's product line for the mining, aggregate quarrying, and construction industries, as well as for the oil industry.

This ad for Gardner Denver tools promoted dependable, easy-to-use tools that enabled operators to safely turn out more work with fewer rejects.

Wire-wrap machine used in the electronics industry

A Successful First Century

Gardner-Denver closed in on its centennial celebration as a successful manufacturing firm in mid-America, strong in its market niche of supplying pumps and compressors to heavy industry and the petroleum sector of the economy, and making and selling rock drilling machinery for the construction and mining industries. The company was well regarded in industrial circles, and it had proved its business acumen at mid-decade when it diversified its product line through the acquisition of Keller Tools and Mayhew Machine.

Conservatively run, Gardner-Denver gained a reputation for reinvesting capital in improving the company's operations. And the postwar years had been very good for the firm. Gardner-Denver revenues had been less than $15 million in the immediate postwar years of 1945 and 1946 and had then plateaued at around $22 million in the late 1940s. Revenues had reach $33 million in 1952 and 1953. But the ramp-up in petroleum exploration in the mid-1950s and the promise of sharply increased highway construction spending after 1956 to begin the interstate highway network was an absolute spur to Gardner-Denver sales. Revenues were just under $48 million in 1955, increasing dramatically to $60 million in 1956 and $72 million in 1957. Only in the recession year of 1958 did sales fall back, and then to $63.5 million, well above the ten-year average of the 1950s.

Mayhew water well rig

The Centennial Year

The Gardner-Denver centennial in 1959 almost never happened. In 1956, new president Gifford Leece entered into lengthy merger negotiations with Dallas-based Dresser Industries. Leece's decision to pursue the merger was a function of his determination to grow Gardner-Denver's business of supplying pumps and compressors to the oil-drilling segment of the petroleum industry.

Dresser was the giant in the oil drilling supply industry in 1956 with more than $200 million in revenues. Leece was concerned that Gardner-Denver's line of oil pumps wasn't enough to compete in the cutthroat world of oil drilling equipment. He was also worried that the Gardner-Denver line of oil pumps would expose the firm to a hostile takeover attempt by some of the larger competitors in the field, including the Continental Emsco Division of Youngstown Sheet & Tube Company, the Oil Well Supply Division of U.S. Steel Corporation, and the National Supply Division of Armco Steel Corporation.

But Leece and Dresser management could never agree on terms for a merger that would treat Gardner-Denver on an equal basis with the larger firm. As a result, Leece made a bold move to broaden Gardner-Denver's offering of pumps and compressors to include a complete line of oil well drilling equipment. In early 1958, Gardner-Denver assigned twenty engineers and technicians to a team responsible for designing new drilling equipment. Late in the year, Gardner-Denver introduced a 70,000-pound draw-works and a 20,000-pound compound power transmission for oil drillers.

To bolster Gardner-Denver's line of oil drilling equipment, Leece made a successful 1957 bid for Mayhew Machine Company and Mayhew Supply Company for a $5 million stock swap. The company made an immediate infusion of capital into the Dallas-based subsidiary and began assembling air-drilling compressors for areas such as Saudi Arabia and the Middle East where water for mud pumps was scarce to nonexistent.

Leece's initiatives to push Gardner-Denver into a major supplier of oil drilling equipment were reflected in the firm's bottom line. Edgar Schaefer, Leece's predecessor as president, presided over a 50 percent increase in Gardner-Denver sales between 1949 and 1954. Leece doubled Gardner-Denver sales between 1954 and 1959, his first five years as president.

CHAPTER 5
Years of Growth and Acquisition

Gardner-Denver continued its expansion program in the 1960s, purchasing the Apex Machine and Tool Company in Dayton, Ohio, to complement the company's growing line of accessories for air compressor equipment. Apex was a leading manufacturer of screwdriver bits and sockets for power tools and an important producer of universal joints for the machine tool and aircraft industries.

Sales increased to a record $85.8 million in 1961, and net income increased slightly to $7.2 million. Even though net income was reduced by sharply increased federal taxes initiated by the incoming administration of President John F. Kennedy, Gardner-Denver continued its tradition of planning for the future by making the capital investments that allowed the company to continue growing. In 1961, Gardner-Denver spent almost $1.9 million to purchase land in Harvey, Louisiana, and Oklahoma City, Oklahoma, for proposed office and warehouse facilities.

Gardner-Denver also continued to innovate. By 1961, one of every fourteen employees was involved in the company's research and development efforts. In the late 1950s and early 1960s, Gardner-Denver's Keller Tools division in Grand Haven, Michigan, became a leader in the manufacture of complex assembly machines for the production of the first generation of computers used by American industry.

Gardner-Denver's wire-wrap machine was a high-speed, numerically controlled device that automatically wired electronic panels. Operating from punched cards, the "wire-wrap" machine was capable of attaching 750 spaghetti-thin wires to a twenty-by-thirty-inch panel in two-and-a-half hours.

Gardner-Denver introduced the Roto-Screw in the early sixties.

Other new products introduced during 1960 and 1961 included the company's line of Roto-Screw compressors, a feed leg drill, and various pneumatic tools. Another new product introduced by Gardner-Denver during the period was a new model compressor developed specifically to drive pneumatic unloading of bulk cement from large mixer trucks. Gardner-Denver made many of the compressor parts for the cement trucks from aluminum, achieving a weight savings of 420 pounds, which then allowed the operator of the truck to carry that much more cement in payload.

Top: Portable compressors
Bottom: Gardner-Denver equipment is used to unload dry bulk from trucks.

A Community Mainstay

Gardner-Denver was one of the mainstays of the local Quincy economy. The company operated two major production plants in Quincy, one on Gardner Expressway and the second in Walton Heights, northeast of downtown. In 1961, the Gardner Expressway plant consisted of a pattern shop, sales office, metallurgical laboratory, research and development facility, and the company's general offices. The Walton Heights plant consisted of a machine shop, as well as assembly, testing, welding, and sheet metal fabricating units.

All told, Gardner-Denver employed 1,710 people in Quincy. The company's 1961 payroll totaled $9.7 million, and the company spent an additional $1.8 million each year in the local area for services and supplies. Gardner-Denver paid $280,000 each year in Illinois state and county taxes.

Finally, Gardner-Denver provided innumerable intangibles to its host community. The executives, toolmakers, machinists, engineers, metallurgists, welders, and assembly workers who made the blue-colored machinery known throughout the world were the backbone of the local community. They gave their time, talents, and money to hundreds of worthy local causes. Gardner-Denver employees could be found in every Quincy community and church organization, from Girl Scouts to Rotary to youth sports activities. And the budgets of most Adams County charities would have been considerably leaner without the contributions from Gardner-Denver employees and retirees.

When the whistle blows

Since the early years, Gardner-Denver has held company picnics at local parks, like these photos taken at Eagles Alps in the 1940s.

Throughout the years, Gardner-Denver has sponsored various sporting activities.

A Gardner-Denver employee provides leadership to a local scout group.

The 1939 Gardner-Denver championship team

Gardner-Denver sales representatives

The Sales Force

Gardner-Denver's success in the 1960s was attributable to the quality of the equipment that bore the company label. But Gardner-Denver also was recognized for the expertise and helpfulness of its sales force.

A job with Gardner-Denver's sales force was a status symbol for young people from Quincy and the local area. To be a local salesperson for Gardner-Denver meant dressing up, and for men, that meant wearing a tie every day. A company car in the 1970s was a Ford Falcon. The road system wasn't what it is today. There was no interstate from Quincy to Springfield. The sales personnel drove Illinois 104 to Jacksonville to U.S. 36 to call on customers in eastern Illinois and neighboring Indiana. Many of the roads taken were two lanes back then. Sales personnel were typically on the road from Tuesday morning to Friday night.

Lodgings were less than deluxe. Most of the communities a Quincy salesperson would service had downtown hotels. Staying at the new Holiday Inn in Peoria for $25 a night was considered a luxury. There were no chain restaurants in Decatur, and Sculley's Chili Parlor and the Coney Island were noted for a bowl of Springfield's signature dish. Contact with the home office was sporadic at best. There were no cell phones, computers were still a generation in the future, and fax machines were only just beginning to make their appearance in corporate offices. The office would call the customer's office if it had to contact a salesperson. The office knew where you were supposed to be at all times. Sales personnel were the public face of Gardner-Denver and a big reason for the company's success. Extremely knowledgeable and focused on the customer's every need, Gardner-Denver's sales staff was a critical component of the company's growth.

Changes at the Top

By 1963, Gifford Leece had been at the helm of Gardner-Denver for nine years. Leece had informed the board in 1962 that he intended to retire in the spring of 1963. Ralph Gardner, the grandson of founder Robert W. Gardner, agreed to vacate his chairmanship to make room for Leece in the position, who moved up to chairman of the board on April 30, 1963.

Like Leece and his predecessor, H. Guy Myers, the new president of Gardner-Denver had come up the ranks on the sales side of the business. Cedric H. Rieman joined Gardner-Denver as a salesman in 1937 and spent thirteen years in sales in Quincy, San Francisco, and Duluth, Minnesota. A native of Vergas in Minnesota's Otter Tail County, Rieman graduated from the School of Mines at the University of Minnesota. In 1950, he was named sales manager of Gardner-Denver's Mining and Construction Division.

Cedric Rieman

After five years as sales manager, Rieman headed west to Colorado where he served as vice president and general manager of the company's Denver plant. In 1961, the board named Rieman executive vice president and asked him to return to Quincy headquarters. Two years later, he capped his twenty-seven-year rise through the ranks by being named Gardner-Denver's seventh president.

Rieman pledged to continue the policies of Leece, his predecessor. He told shareholders in early 1964 that the record sales and earnings achieved during his first year in office "are a reflection of the dedicated service and the cooperation extended by our employees, shareholders, suppliers and friends. The continuation of this cooperative effort in serving our customers' needs will be carried forward. We will, as in the past, emphasize product improvement and operating economies."

Rieman and Leece lost one of Gardner-Denver's most engaging ambassadors when Ralph G. Gardner died in November 1964. The only son of John Willis and Helen Gifford Gardner, and the grandson of Robert W. Gardner, Ralph Gardner was born in Quincy in 1890. A graduate of Purdue University, he returned to Quincy in 1911 to join his father in the operation of Gardner Governor Company. When Gardner Governor and the Denver Rock Drill Manufacturing Company merged in 1927, Ralph Gardner was named vice president and treasurer of Gardner-Denver, a position he held until 1950 when he was elected chairman of the board.

Ralph Gardner

Already in ill health in 1963, Gardner stepped aside to make room for Leece to become chairman of the company. During his fifty-plus years with Gardner-Denver, Ralph Gardner had been a tangible link to his grandfather, the founder of the company. He also was a pillar of the Quincy community, helping establish the Quincy Park District as a model for community beautification and recreation. His donation of a family hunting preserve on the Mississippi River north of Quincy helped establish the Gardner Unit of the series of Mark Twain Wildlife Refuges along the river.

The death of Ralph Gardner severed the company's most visible link to the past. But the entrepreneurial spirit of the Gardner family lived on in the commitment of thousands of Gardner-Denver employees to make the Quincy company a leader in American industry.

Five Years of Record Sales and Profits

Gardner-Denver continued its record of increasing sales and profits as the decade of the 1960s reached its halfway point. In 1965, sales and dollar earnings rose to record levels for a fifth consecutive year. Sales of $123.3 million were up 15 percent over 1964 sales, while earnings of just over $15 million exceeded those of 1964 by 22 percent. Earnings per share of $2.85 compared with $2.32 in 1964.

As usual, the company continued to invest in capital improvements that would be reflected in the bottom line in years to come. In 1965, Gardner-Denver spent almost $4.3 million on factory additions and improvements.

Gardner-Denver will acquire Martin-Decker

Gardner-Denver followed up on the 1961 acquisition of Apex Machine and Tool Company with the 1966 purchase of the Martin-Decker Corporation of Santa Ana, California. The company exchanged 150,000 shares of Gardner-Denver common stock for all of the Martin-Decker capital stock outstanding. Martin-Decker was one of the leading manufacturers of recording instruments for oil drilling rigs, as well as weighing devices for the petroleum industry. The Martin-Decker acquisition gave Gardner-Denver a more complete package of equipment available to its petroleum customers.

Gardner-Denver's financial strength was reflected in the increase of the company's equity. In early 1963, stockholders approved a 3 for 2 split of common stock, and the next year, directors approved a jump in the common stock dividend from 37.5 cents a share to 43.75 cents a share. The company's stock price of $33 a share in the mid-1960s was below the record $38 a share it reached in 1959 and again in 1961, but was still strong when adjusted for the 1963 stock split.

In April 1965, the board again voted to split the stock 3 for 2 and approved a 32.5 cents dividend on the adjusted shares. The third and final stock split of the decade came in early 1969 when common shares were split 3 for 2. The board voted to continue the 32.5 cents dividend on the adjusted shares.

Beefing Up Mining and Construction

The Apex Tool and Martin-Decker acquisitions positioned Gardner-Denver to be a strong competitor in the petroleum equipment markets in the second half of the 1960s. In line with its longtime strategy of diversification, the company announced early in 1963 its plans to increase its offerings for the mining and construction segments of its business.

"We're putting our heaviest bets on introducing large quarrying and drilling machines for the mining, construction and concrete aggregate industries," Chairman of the Board G. V. Leece told New York financial reporters, "We expect to come out with quite a few of these in the first quarter, and by the third quarter they should be picking up some momentum."

Gardner-Denver's Denver factory complex had been expanded and improved in the early 1960s to handle the expected increase in drilling and quarrying equipment. In 1961, crews demolished the 120-foot dual smokestacks that towered over the Denver site for more than forty years. That same year, Gardner-Denver completed a 240-foot-long steel

Gardner-Denver portable compressor and drill operating at a quarry

storage area with a five-ton bridge crane. The company built a new steel processing building at the Plant 2 site and used more than three thousand gallons of paint to give the Denver complex an entirely new look.

The next year, crews completed a new receiving building at the north end of the Plant 2 facilities. Two stories high with a four-ton elevator, the new receiving building was fully outfitted with 440-volt power. Improvements in 1963 included all new machine tools, and the following year, the Denver Division completed a six-thousand-square-foot addition to the KM building. The company erected a $35,000 substation in 1965 to extend 440-volt power to the entire complex.

Gardner-Denver's capital expenditures paid dividends as the company's Denver Division built such mammoth drilling equipment as "Crustbreakers" for the aluminum smelting industry and the new PR143J drilling machine with its specially designed bit holding four large slugs of tungsten carbide.

Expansion continued throughout the latter half of the 1960s. In 1966, the Denver Division entered the computer age with the installation of an IBM 1030 system. Crews swarmed across the complex, building three new structures at Plant 9 to enlarge Department 45 for the installation of new turret lathes, a two-floor office building and a new maintenance department section. Contractors completely remodeled the Personnel Building at Plant 2 and made extensive improvements in the Heat Treating and Plating section of the complex.

The increase in parts manufacture at Denver Division necessitated the purchase of twenty new machine tools in 1967, and 1968 saw the construction of a 28,000-square-foot expansion at Plant 9, which allowed managers to move the Mills and Drills Department from Plant 2, freeing up some desperately needed floor space at the older plant. An L-shaped building expansion at Plant 9 in 1969 created

Four boom truck jumbo

space for the Receiving Department, installation of an Inertia Bonding Machine and storage for the finishing of aluminum guide shells, and conversion of the second floor of the office space as a conference room and company credit union. A 1971 expansion at Plant 9 provided ample room for an enlarged steel storage system, inspection department, and maintenance department. The company spent $700,000 that year to upgrade the Denver Division with new, faster machine tools.

Gardner-Denver also strengthened the manufacturing capacity of the Denver Division with expansions away from the Denver complex. Gardner-Denver purchased a ready-made factory in Coffeyville, Kansas, for future expansion. The firm purchased Funk Manufacturing Company in South Coffeyville, Oklahoma, in 1967, a respected manufacturer of gears and power trains.

Denver Division's strong sales during the 1960s contributed to Gardner-Denver's success during the decade. The Denver Division more than doubled in size during the decade, as Gardner-Denver made strong inroads into the construction market, particularly in providing the equipment needed by contractors to build America's ambitious interstate highway system.

Into the 1970s

Gardner-Denver finished the 1960s with a flourish. The decade ended much better than it had begun. Sales and earnings in 1969 were at an all-time high. Sales of $193.7 million were 12.4 percent higher than 1968. Profits of $23 million, or $2.85 a share, were 17.5 percent higher than the previous year. Dividends increased to $1.30 a share following the 3 for 2 stock split in January 1969. It was the sixth dividend increase in seven years.

As in previous years, Gardner-Denver continued to invest in capital expenditures for the future. The 1969 expansions and plant improvements included a host of projects. Gardner-Denver opened a new sales office, service, and minor assembly facility at its Martin-Decker subsidiary in Houston, Texas, and new office and warehouse facilities for its operations in Bakersfield, California.

With its lines of equipment for the petroleum, construction, mining, and quarrying industries, Gardner-Denver pronounced itself "truly in a position for assured growth." By 1969, the company had identified the growth segments of the economy for the 1970s, including computerization, air and water environmental mitigation, communications, and aerospace technology.

A "world catastrophe," however, was waiting in the wings. Global energy shortages in the 1970s would test Gardner-Denver management as never before.

The Mississippi River Flood

The flood of 1929. *Courtesy of the Historical Society of Quincy and Adams County*

Every generation of Gardner-Denver employees has flood stories to tell. The Mississippi River flows past Gardner-Denver's front door in Quincy, Illinois, and the river has an unseemly habit of spilling out of its banks on a pretty frequent basis.

In 1927, at about the time that Gardner Governor completed its merger with the Denver Rock Drill Manufacturing Company, the Mississippi unleashed some of the worst flooding in its history all the way from Minnesota and Wisconsin to the Gulf of Mexico. The rains had started falling across the Upper Midwest in August 1926, and they continued for much of the fall and winter. The Quincy waterfront and much of western Illinois and eastern Missouri were under water for months at a time.

As the crest rolled south from the confluence of the Ohio at Cairo, Illinois, the flooding got worse and worse. By the time it reached Louisiana in the spring of 1927, more than one million people in the flat bottomlands of the Delta were left homeless.

Other floods along the Mississippi were memorable for Gardner-Denver employees and Quincy residents. The floods of 1973, 1993, and 2008 are all remembered for their destructive force. But perhaps the most miserable flood in Quincy's history was the Mississippi rampage of March and April 1965.

"Only one thing is worse than wrestling with a berserk river," wrote a chronicler of the 1965 flood, "wrestling with a freezing cold, ice-choked berserk river." Heavy snow in Minnesota and Wisconsin during late 1964 and early 1965, coupled with heavy rains in the Upper Mississippi Basin in February and March 1965 sent the river barreling southward at flood stage. By the time the swollen Mississippi reached the Quad Cities above Quincy on April 28, it was at a record crest of twenty-two-and-a-half feet, three feet above the 1892 record.

With nowhere to go, the river spilled over into the floodplain for miles, inundating hundreds of square miles between the Quad Cities and Quincy. Before it was over, the flood would cause more than $225 million in damages to public and private property along the river between St. Paul and Quincy.

The Gardner Expressway plant was hit extremely hard, with water filling in the area around the ice plant and much of the area on the south side of Curtis Creek under water.

Aerial shot of the Quincy plant during the 1973 flood

In meetings with the staff of the Rock Island District of the U.S. Army Corps of Engineers a year after the April 1965 flood, executives at Gardner-Denver received a briefing on the Corps' plans for flood control for the Upper Mississippi. Still, the engineers were blunt in their assessment of the Quincy Plant and the LaGrange Foundry in the face of future floods.

Levees at both facilities should be raised to thirty feet, the Corps recommended, and Gardner-Denver should seriously consider future expansion well away from the river. "Any long-range expansion programs should be located in another site, such as the Walton Heights Area," the Corps' report said.

Scene from the 1973 flood

The head of the U.S. Army Corps of Engineers, General Robert Van Antwerp, came to Quincy to personally view the work done by the employees in their fight to stay dry during the flood of 2008.

While the employees fought the flood of 2008, production didn't miss a beat as drilling pumps continued to ship out.

Wes Carl and Cindy Frericks fill one of the 400,000 sandbags used in 2008.

Phil Schults surveys the river level in 2008.

CHAPTER 6
Riding the Boom, Riding the Bust

After more than a quarter-century of prosperity, the United States economy encountered hard times during the late 1960s. The Arab Oil Embargo in the third week of October 1973 only made a bad situation worse. As Israeli Defense Forces flanked the Bar Lev Line and re-took the Suez Canal in the short but brutal Yom Kippur War, Arab oil sheikhs protested American support of Israel by embargoing Middle Eastern oil shipments to the United States and its Western European allies.

The administration of President Richard M. Nixon was caught off guard, reeling from the October resignation of Vice President Spiro Agnew on alleged corruption charges. Nixon also was struggling with efforts to contain the rapidly escalating Watergate Crisis, which culminated that week in presidential agreement to turn over the Watergate tapes and Nixon's almost immediate sacking of special prosecutor Archibald Cox, followed by the resignation of top Justice Department officials.

As a result, Nixon and the nation were caught by surprise when gas prices began to rise in November. By the end of the year, gas at the pump had doubled and would double again during the first quarter of 1974. Federal standby rationing plans spooked many motorists, and Quincy and many other communities were soon the scene of long lines at gas stations.

The oil crisis of 1973 was quickly followed by rapidly escalating inflation and skyrocketing interest rates. America had tried to fight a war in Vietnam in the late 1960s and early 1970s while maintaining a peacetime economy. The result was chaos in the financial markets by the mid-1970s. Jimmy Carter, the president of the United States from 1977 to 1981, watched helplessly as short-term interest rates ballooned to nearly 20 percent. Inflation ravaged the economy, with products from milk to meat to motor oil increasing in price as much as 10 percent a year and more from 1977 to 1981.

Gardner-Denver equipment is used on offshore rigs around the world.

The impact of rising interest rates and inflationary pressures led to labor discontent in the United States. Unions, whipsawed by grass roots dissatisfaction with prices and bank interest rates that were lessening the value of a paycheck almost on a weekly basis, became increasingly militant.

For Gardner-Denver, as for much of the American manufacturing economy, the 1970s were a difficult decade. Corporate expansion slowed appreciably during the late 1960s and early 1970s, and Gardner-Denver was forced into a more conservative focus than it had enjoyed during the 1950s and 1960s.

But Gardner-Denver's reputation for quality in its line of pumps, compressors, and drilling equipment ensured that the company would continue to maintain its decades-long record of products. In fact, between 1966 and 1975, while much of America's manufacturing industry was struggling to stay in the black, Gardner-Denver consistently operated with a profit margin of 21 percent or higher.

A New Leader

Along with the old, Gardner-Denver celebrated the new in 1973. After more than thirty-five years of service, Cedric Rieman announced his intention in early 1973 to step down as Gardner-Denver's president and chief executive officer, a position he had held since 1963. "The Company fortunately has young and capable managers ready to assume the responsibility of operating the business," Rieman told shareholders in announcing his decision.

Rieman's successor was one of those capable young managers, Lynn Leigh. Then forty-three, Leigh joined Gardner-Denver as administrative assistant to the general manager of the company's Dallas Division in 1967. An Idaho native and a graduate of the University of Kansas with a degree in aeronautical engineering, Leigh had served in the U.S. Navy during World War II and had been employed in sales, manufacturing, and general management positions with firms in the oil field machinery business for twenty years prior to joining Gardner-Denver.

Lynn Leigh

Leigh was later named assistant manager of the Dallas Division and in April 1971 was elected to the company's board of directors. Less than a year later, in the spring of 1972, Leigh was asked to move to Quincy as vice president of operations.

The company that Leigh assumed executive responsibility for in early 1973 was a diversified manufacturer with almost $210 million in sales. Net income of $23 million resulted in earnings per common share of $1.39. And the company board increased the dividend to 17.5 cents a share in early 1973.

Gardner-Denver's lead in mining and quarrying equipment was undisputed in 1973. The huge new GD-120 and GD-130 rotary drilling rigs introduced in the early 1970s were gaining an enthusiastic reception in coal, metals, and quarrying applications. Self-contained secondary drilling units that combined a crawler-carrier, air compressor, drill, and feed were developed for open pit mines in the United States and abroad.

New products combined with broadened applications guaranteed that Gardner-Denver equipment found increasing employment in a wide range of applications in the manufacturing, transportation, food and beverage, and energy sectors.

In petroleum, Gardner-Denver remained strong. "Drilling units that include Gardner-Denver equipment are in use everywhere in the free world where oil and gas are found—the Middle East, Far East, Europe, North Africa and South America, as well as all domestic fields and offshore locations," the company told shareholders in its annual report.

Gardner-Denver's growing presence in international operations marked the company's emergence as a global competitor. Gardner-Denver equipment was utilized in Saudi oil fields, tunneling projects in Japan, a large hydroelectric dam in Afghanistan, copper mining in New Guinea, a Ford plant in Germany, water and wastewater treatment plants in West Africa, open pit iron mines in Brazil, and railroad projects in Canada. Indeed, by 1973, almost 40 percent of Gardner-Denver's revenues were derived from overseas sales. And Gardner-Denver's international presence was further enhanced in 1973 when the company reached out to acquire a trio of British companies.

Cable binding tool

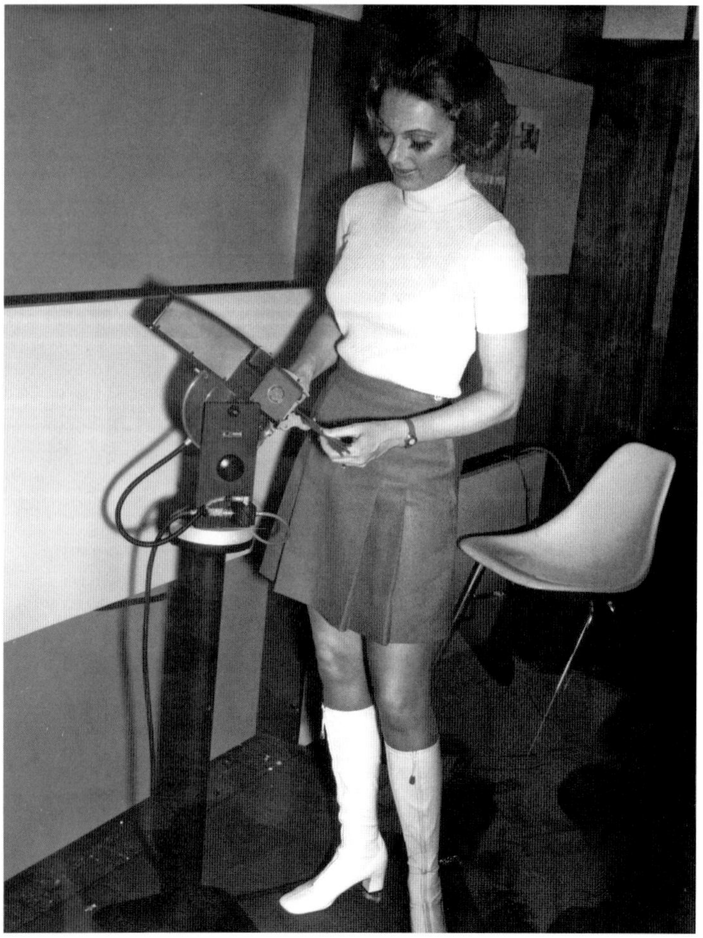

Crossing the Atlantic

America had fallen in love with British music and culture in the late 1960s and early 1970s. The Beatles, Rolling Stones, Byrds, and other British groups transformed American popular music, and British actors like Peter Sellers, Sean Connery, Richard Harris, Peter O'Toole, Richard Burton, and Michael Caine dominated the screen on the American side of the Atlantic.

So it was only natural that Gardner-Denver would be attracted to British companies when it embarked on an initiative to begin acquiring companies overseas. The first acquisition was announced in the fall of 1972 when Gardner-Denver acquired two manufacturing units from the British firm Tarmac Ltd. The first two acquisitions were Padley & Venables Ltd., a producer of coal picks and accessories for rock drills, and Bedford Steels Ltd., a producer of tool steel for forming into drill rods for use with rock drills. At the same time, the company acquired the full ownership of Bedford Rock Drill Components Company, a subsidiary it had owned with Tarmac Ltd. since 1972.

Padley & Venables was founded in 1911 as R. Padley & Company in a small factory at Hill and Boston streets in the industrial heart of Sheffield. Thomas E. Venables joined Padley in 1918, and the partnership was formalized as Padley & Venables two years later. During the 1920s and 1930s, the company manufactured hand tools, factored one-piece drill steels, pneumatic chisels, rivet snaps, and coal-cutting picks. Following World War II, Padley & Venables Ltd. became a public limited company in 1947.

The postwar years saw the newly public company undertake a major expansion. In 1956, Padley & Venables moved from Sheffield to a new, nine-acre complex at Dronfield, a suburb in the nearby county of Derbyshire. Four years later, the company acquired Production Tool Alloy Company Ltd. of Bedford, which gave Padley & Venables capabilities in stainless steel production, tungsten carbide manufacturing, and operations overseas. In 1962, Derbyshire Stone Ltd., a nearby large quarry operation, acquired Padley & Venables as part of a diversification initiative.

Padley & Venables went through one final acquisition before becoming a part of the Gardner-Denver family. In 1968, Derbyshire Stone Ltd., the firm's parent, merged with Tarmac Ltd.

Bedford Steels Ltd. was another Sheffield-based industrial company whose origins traced to the World War I era. In 1918, just prior to the armistice ending the Great War, John Bedford & Sons Ltd. was granted a patent specification for the rolling of a hollow mining drill steel. Mining customers had long been asking for the production of a hollow drill steel, allowing the bit to be flushed with water.

John Bedford & Sons assigned the patent to Bedford Steels Ltd, the subsidiary set up to manufacture the new hollow drill steel. By the early 1920s, the new company was established on Effingham Road in Sheffield, adjacent to the Yorkshire Navigation Company Canal. The Bedford Steels rolling mill was equipped with a two-high twelve-inch mill, a nine-inch mill, and an eight-inch mill. The plant site on the canal allowed the company to use Swedish carbon steel billets imported directly to the mill by water via the east coast port of Goole.

Bedford Steels replaced the original twelve-inch mill with a new Double-Duo mill in 1958 and considerably extended the firm's ancillary equipment and warehousing space in the 1960s. In 1970, John Bedford & Sons Ltd., the parent company, was acquired by Tarmac Ltd., bringing the Bedford Steels rolling mill under the same ownership umbrella as Padley & Venables. In June 1972, Tarmac sold the hand tools division of John Bedford & Sons to Spear & Jackson International Ltd. and the following year sold Bedford Steels Ltd., along with Padley & Venables, to Gardner-Denver.

The 1973 acquisition of the British firms immediately gave Gardner-Denver a strong corporate presence on the European Continent. The firm had long operated sales and manufacturing facilities to serve local mining customers, but the purchase of Bedford Steels and Padley & Venables was seen as part of a global corporate expansion strategy.

Drawworks used in oil drilling

Even before the deal was closed, Gardner-Denver President Lynn Leigh announced that the company was opening a separate office to oversee the firm's international operations. What raised eyebrows among members of the Gardner-Denver community was where the new international office would be located. Early in 1973, Leigh announced that the new office would open in Dallas, Texas, instead of Quincy.

A New Image

The 1970s saw substantial growth. Gardner-Denver reported a sales increase of 30 percent in 1973 and a gain in earnings per share of 20 percent. Gardner Denver supplied the coal mining and petroleum industries with a wide range of equipment, and the energy shortage kept Gardner-Denver very busy in that area. With total foreign sales bringing in more than one-third of the company's revenues, Gardner-

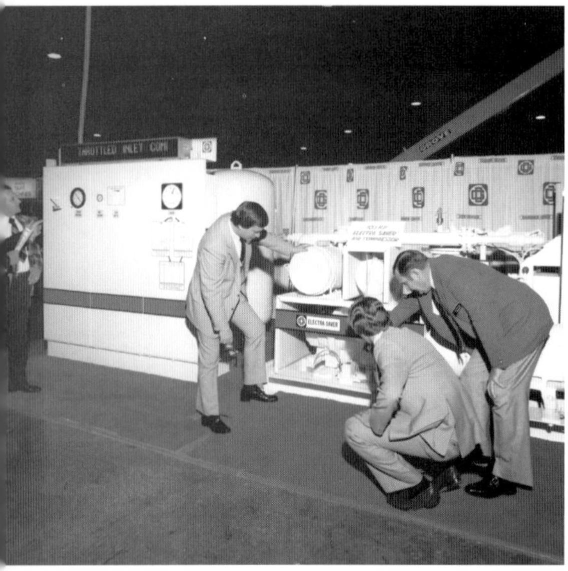

Gardner-Denver introduced the Electra Saver Turn Valve compressor in 1975, setting a new bar for energy-efficient, rotary screw compressors.

Denver was well placed to benefit from increased globalization.

In the early 1970s, mining, for example, shifted from deep shaft boring in which Gardner-Denver drill rigs were indispensable to open pit excavation, a market for which the company had much less to offer. Lynn Leigh resolved to do what it took to position Gardner-Denver for growth during the second half of the 1970s.

Finishing Strong

Gardner-Denver appeared to finish the decade of the 1970s on a high note. Leigh's strategy of growing the company's sales first and worrying about profit margins later was highly successful. Revenues were $355 million in 1974, $423 million in 1975, and $443 million in 1976. And revenues really picked up the next two years, topping $500 million in 1977 and reaching $652 million in 1978.

Profits, however, failed to keep pace with revenue growth. After reaching $33 million in net income in 1975, earnings in 1976 and 1977 of $20 million and $23.6 million reduced profit margin to less than 5 percent.

Gardner-Denver went on to report one of its best years in history. The company booked sales of $650 million in 1978, a compound annual rate of growth of 16 percent during the previous five years. International sales increased 22 percent during the year, accounting for one-third of total sales. More importantly, 86 percent of Gardner-Denver products sold worldwide were manufactured in the United States.

The best news of all in 1978, however, was the strong growth of profit. Net income for the year was $51.4 million, more than double the net income booked in 1977. As a result of the strong growth in earnings, the board of directors raised the dividend twice during the year, to twenty-one cents a quarter in February and to twenty-five cents a quarter in October. The $1 annual dividend rate at year-end represented a whopping 32 percent increase over the dividend paid in the previous year.

On January 22, 1979, the boards of directors of Gardner-Denver and Houston-based Cooper Industries Inc. made a joint announcement that both companies had approved a definitive agreement providing for the merger of Gardner-Denver into Cooper.

Gardner-Denver sales representatives

Various logos used during the first one hundred years of the company

Gardner-Denver's drawworks were used at oil rigs around the world.

Workers in the Quincy plant

The Move to Dallas

No sooner had Gardner-Denver opened its new international office in Dallas than Lynn Leigh approached the board of directors with a proposal to move the corporate headquarters from Quincy to Dallas. Leigh, who had longtime roots in the Texas community, noted that a large amount of the company's business was with Texas petroleum interests, and that long-term growth prospects for the company's petroleum lines was excellent. Furthermore, Gardner-Denver at the time was in the process of expanding its manufacturing facilities in Dallas.

In June 1973, Gardner-Denver's board of directors approved the move of corporate headquarters from Quincy to Dallas. Leigh, the executive team and approximately thirty-five other employees and their families would be moved to Texas beginning in the fall of 1973. Initial plans called for sharing office space with the international division, and the target date for full operation of the new offices was scheduled for mid-1974.

The move to Dallas wasn't universally accepted. In 1973, Gardner-Denver had been a part of the Quincy community for 114 years, and generations of local residents held stock in the hometown company. Soon after the board of directors announced the move of the corporate headquarters, a group of shareholders organized the Gardner-Denver Stockholders Defense Group to oppose the move. Prominent in the group's leadership were Gifford Leece and Peter F. Fleischmann.

Leece, a former Gardner-Denver president, CEO, and chairman, was a longtime Quincy civic leader. Fleischman, the grandson of J. Willis Gardner, was president and chairman of the board of the *New Yorker* magazine. Between them, Leece and Fleischman controlled 15.4 million shares of stock. The Gardner-Denver Stockholders Defense Group retained the Chicago law firm of Kirkland and Ellis to assist it in convincing a majority of shareholders to oppose the move to Dallas.

Leece, Fleischmann, and the stockholders group rebutted just about all of the arguments for moving made by Leigh and the board of directors. They pointed out that Quincy had scheduled air service, and that Gardner-Denver's two aircraft allowed executives to make air connections in St. Louis a half-hour away, and in Chicago, in less than an hour. Furthermore, the opponents claimed, Gardner-Denver's international business could be better served from Quincy since Chicago was much more of an international air hub than Dallas.

In the end, Gardner-Denver did move its corporate headquarters to Dallas in 1974, but the vast bulk of the company's Quincy employees stayed put. The stockholder group's opposition eventually subsided, and Quincy remained an integral part of the Gardner-Denver culture.

But Leigh and the board of directors did use the Dallas move to bring new blood into top management. In 1975, Gardner-Denver announced the appointment of four new directors, all with extensive management experience in American industry. John C. "Jack" Heisler was named Gardner-Denver's financial vice president and treasurer. A trio of outside directors, Peter Scotese, Stanley E. G. Hillman, and Harker Collins, joined Heisler on the board.

Gardner-Denver Corporate Headquarters in Dallas, Texas

Peter Scotese, a New York City resident, was president and a director of Springs Mills Inc., one of the nation's best-known textile firms. Scotese also served as a director at Bell & Howell. Stanley E. G. Hillman brought a lifetime of international experience to Gardner-Denver. The Chicago resident served as president, director, and chief operating officer of IC Industries, a diversified holding company with fifty-nine plants in the United States, Canada, Mexico, and Europe. Hillman, a graduate of the London School of Economics, spent the first twenty years of his business career in Great Britain, Shanghai, and Hong Kong.

Harker Collins, the president of Bandag Inc., was familiar with the problems of running an industrial company in a smaller midwestern community. Bandag, a manufacturer of recapped tires for trucks and heavy equipment was headquartered in Muscatine, Iowa, just upriver from Quincy.

Leece continued to oppose Leigh's management team. In 1977, Leece spearheaded a new group, the Gardner-Denver Company Stockholders Defense Committee. At the 1977 annual meeting of stockholders, the committee unsuccessfully nominated J. W. Gardner II, a former director, and R. H. Myers, a Quincy high school history teacher, to serve on the board of directors and bring the management team back to Quincy.

CHAPTER 7

The Cooper Industries Years

The events that brought Gardner-Denver together with Cooper Industries had been building for a number of years. The 1970s had been a boom period for the capital goods sector of the American economy, and Gardner-Denver had ridden that boom to record levels of prosperity. But the company's expansion and its growth from $200 million of annual sales in the early 1970s to $650 million in 1978 exacted a steep price.

In the 1960s, Gardner-Denver had been able to maintain profit margins of 20 percent and better. But once the company started growing, profit margins dropped precipitously. By 1975 and 1976, the company's profit margin was flirting with 5 percent.

Wall Street's expectations were that a company with a track record of profitability should be able to maintain that record of profits indefinitely. And if the company couldn't perform, Wall Street would help it be acquired by a company that could. *Forbes* magazine, for one, was very critical of Gardner-Denver, noting the Quincy firm was "a company notorious for lack of planning or cost controls."

The 1970s was a difficult period for any company to make an impression on the Wall Street investment community.

The Dow Jones Index suffered through a decade-long bear market, and when the price of trying to fight a war in Vietnam with a civilian economy collided with the first Arab Oil Embargo of 1973, the economy was set up for an even more difficult period.

So when Cooper Industries approached Gardner-Denver during the fourth quarter of 1978, Lynn Leigh and the board of directors were receptive to the idea of a merger. Wall Street was enamored of economies of scale and rewarded companies that could both get bigger and remain profitable. Leigh put it best when he explained to shareholders that Cooper was "a leading manufacturer of hand tools, a principal independent supplier of maintenance and repair services for aircraft and aircraft engines, and a major producer of engines, turbines and gas compressors for the world's oil and gas markets. We believe that the strengths of the two organizations complement each other, and the combined company will offer substantial potential for superior performance."

Early Gardner-Denver portable compressor

Like Gardner-Denver, Cooper Industries could trace its roots to the nineteenth century. Begun as a small iron foundry in Mount Vernon, Ohio, by brothers Charles and Elias Cooper in 1833, Cooper Industries initially produced stoves, plows, and other agricultural implements. Renamed the C and E Cooper Company following the Civil War, the firm manufactured steam engines and steam thrashers. In 1869, the company became the first in the Midwest to manufacture the Corliss engine, which had wide adaptation for industrial and power production in the latter half of the nineteenth century. In 1875, Cooper developed a steam power traction engine, which would become America's first engine-propelled farm tractor.

Cooper founder, Charles Cooper

The company stayed in the control of the Cooper family throughout the nineteenth and early twentieth centuries, although Frank Fairchild, the firm's star salesman, was Cooper's first president. Following the turn of the twentieth century, Cooper began to deemphasize Corliss steam engine production and focused on manufacturing natural gas internal combustion engines, used as compressors on early natural gas pipelines in Ohio, West Virginia, and Indiana. During World War I, the firm manufactured hydraulic forging presses for the country's defense effort. Following the war, Cooper merged with the Bessemer Gas Engine Company of Grove City, Pennsylvania, in 1929 to form Cooper-Bessemer Corporation.

COOPER
Gardner-Denver
Industrial Machinery

Under Cooper Industries, Gardner-Denver was called the Industrial Machinery Division.

The new company was devastated by the Great Depression, which began the same week Cooper-Bessemer opened its corporate offices in New York. During the 1930s, the company developed a prototype diesel locomotive, and when America entered into World War II in 1941, Cooper-Bessemer became a major producer of diesel engines for locomotives used by the U.S. Army and for U.S. Navy ships. By the time the war ended in 1945, Cooper-Bessemer employed almost five thousand people.

The postwar years were consumed by a 1950s takeover attempt by a private investment group and by a drop in the firm's pipeline revenues. At the same time, Cooper-Bessemer was growing its international business; at the beginning of the 1960s, the company was strongly established in more than a dozen countries and operated a wholly owned British subsidiary. Cooper-Bessemer's business had branched out into compressors and pumps for oil and gas production, which made the company a direct competitor of Gardner-Denver. In 1965, the company dropped the "Bessemer" from its name and became Cooper Inc.

In 1967, Cooper Inc. moved its corporate offices to Houston and embarked on an acquisition program, which brought a number of well-known tool brands into the Cooper fold, including Lufkin, Crescent Niagara, J. Wiss & Sons Company, and Nicholson File Company. In the 1970s, Cooper began acquiring a number of aircraft services and equipment companies.

In the late 1970s, Cooper capped its decade-long acquisition spree when it acquired Gardner-Denver. Slightly larger in sales volume and assets than Cooper, Cooper had long felt that combining Gardner-Denver's three energy-related business units into its own energy and petroleum division would make sense from an economy of scale perspective. In its first full year of operations, the merged company reported more than $1 billion in sales.

But before the reorganization of Gardner-Denver into Cooper Industries could prove the wisdom of the merger, Cooper and Gardner-Denver faced a recession that rocked the capital goods sector of the North American economy in 1980 and 1981 and rippled into the energy markets beginning in 1983 and 1984.

Horrible Years

In 1978, Gardner-Denver had embarked on a $20 million expansion and modernization program, betting that there would be a strong market in the early 1980s for the company's pumps, drilling equipment, and air compressors. Instead, 1979 and 1980 were horrible years. Iranian students overthrew the Shah and held the entire embassy staff in Tehran hostage for more than a year. President Jimmy Carter stumbled from crisis to crisis.

What would come to be seen as a major restructuring of the U.S. capital goods sector began in late 1979 when U.S. Steel Corporation announced it was closing several obsolete mills. In many ways, the announcement slipped under the radar of the nation's news coverage since America's attention that fall was riveted on the fall of the Shah of Iran and the seizure of the U.S. embassy in Tehran by radical Islamic students.

Portable compressors had long been manufactured in Quincy. Dave Yager and Harry Yackley are pictured in 1980 with the last portable compressor made in Quincy.

But by 1980 and 1981, it was obvious that the problems that caused U.S. Steel to shutter mills were endemic to the integrated segment of the steel industry. Bethlehem, Jones & Laughlin Steel, Armco, LTV, Youngstown Sheet & Tube, and other giants of the industry all announced their own cuts during the early 1980s. Mills and blast furnaces across the Monongahela Valley, South Chicago, Gary, Cleveland, and Ohio's Mahoning Valley closed in 1981 and 1982.

Steel employment nationwide sagged. The Monongahela Valley up and down river from Pittsburgh—long the cradle of American steel—was hit particularly hard. In 1959, more than seventy thousand residents of the Mon Valley worked for Big Steel. By year-end 1981, employment had been slashed in half, to thirty-five thousand steelworkers. Six years later, in 1987, mill employment in the Mon Valley would be four thousand, just over 10 percent of what it had been in 1981.

When the nation's steel industry caught pneumonia, much of the nation's capital goods sector went on life support. Many of Gardner-Denver's customers in the mining, construction, petroleum, and general industrial sectors of the economy shut down plants, laid off workers, and cut back business.

Cooper, which had a reputation for ruthless efficiency, was quick to staunch the bleeding. One of the first casualties was Gardner-Denver's corporate office in Dallas, which was closed in 1979. Employees were transferred to Houston or joined the Construction and Mining Group in Dallas. When the petroleum business began to slow down in early 1982 as oil prices dropped and supplies rose, Cooper instituted major layoffs at the Gardner-Denver operations in Quincy. By June 1982, the company had laid off more than five hundred employees in its Quincy operations. Before the layoffs began, Cooper's Gardner-Denver unit had more than seventeen hundred people on the payroll in Quincy, making the firm the community's largest employer.

Because of the petroleum glut and the ensuing recession, Gardner-Denver's Quincy operations doubled the firm's annual two-week maintenance shutdown to four weeks in the summer of 1982. Even more workers were laid off when they returned to work following the shutdown in July 1982.

Cooper's ruthless cost cutting did produce results. The company reported a record $2.395 billion in revenues in 1982 and actually eked out a profit for the year of $133 million.

The 1982 recession in the capital goods sector was one of the longest running in the post–World War II era, and Gardner-Denver's business did not begin to recover until mid-decade. And when it did recover, the company's parent, Cooper Industries, would be focused on another major merger.

Diversification and Growth

Cooper Industries continued to diversify and expand through acquisitions during the 1980s. In 1981, the company entered a takeover battle for Crouse-Hinds Company as a white knight for the Syracuse, manufacturer of electrical products. The acquisition also brought to the table the Belden Corporation, a wire and cable manufacturer that Crouse-Hinds had been in the process of acquiring when it had to fend off a hostile takeover attempt by InterNorth Corporation.

The Crouse-Hinds and Belden acquisitions were equal in size to the Gardner-Denver acquisition just two years before and increased the size of Cooper Industries by 50 percent. Robert Cizik, Cooper's CEO, explained to financial reporters that the acquisitions were part of Cooper's strategy to diversify its product mix.

That diversification strategy was in the spotlight in 1985 when Cooper followed up its foray into the electrical products business with an acquisition of Chicago-based McGraw Edison Company. McGraw Edison was a major player in the manufacture of electric and lighting products for industrial, commercial, and utility use. Overnight, Cooper doubled in size to a $3 billion company and became one of the world's largest lighting manufacturers.

For Gardner-Denver, the acquisitions by the parent company were important in that it gave Cooper Industries the financial wherewithal to make capital investments in the groups in which Gardner-Denver did business. For the Quincy community, the fear during the recession of the early 1980s was that Cooper would close the Industrial Machinery Division or move it to more modern facilities in Texas.

Those fears were laid to rest in 1983 when Cooper announced it was consolidating operations in Sunnyvale, Texas, and the Ajax Pump line from Corry, Pennsylvania, into the Gardner Expressway plant. The move involved relocating forty-one large pieces of equipment from Sunnyvale to Quincy and pouring twenty new foundations to hold them.

William Lowen, the Quincy plant's production manager, noted that the transfer of the Ajax line would result in the Gardner Expressway plant manufacturing pumps ranging from 150- to 600-horsepower and weighing upwards of fifteen thousand pounds. For the first time, the Quincy plant began manufacturing pumps for saltwater, as well as mud pumps for oil rigs.

Nish Teshoian

Nish Teshoian, vice president and general manager of Cooper's Petroleum Equipment Division, told Quincy reporters that the move is "another positive influence on our business for the future. Our regular product line in Quincy, plus the additions from Sunnyvale and Ajax, will give us the foundation for growth."

That didn't mean that Cooper wouldn't hesitate to consolidate operations and cut costs in Quincy when it saw

Walton Heights was converted to a warehouse in 1985.

the opportunity. In 1985, about the time it was entering into final negotiations to purchase McGraw Edison, Cooper announced it was consolidating operations of its two Quincy plants at the Gardner Expressway facility. Plans called for the company's Walton Heights plant, which Gardner-Denver had acquired from Otis Elevator in the 1940s, to be converted to a warehouse and storage facility. The transfer of 175 jobs and the elimination of about 90 jobs associated with the Walton Heights plant would leave about 780 people employed in Quincy.

The introduction of the M-Series compressor line in 1986 was another shot in the arm to Quincy's Industrial Machinery Division. The first piston compressors manufactured by the division designed to produce oil-free air, the M-Series came on the market just as the manufacturing

and capital goods sectors of the economy began to expand in the wake of the early 1980s recession.

"There is no question it is strengthening our business," Cooper's Teshoian said of the M-Series line. "It is a win-win situation for us and for Quincy."

By the mid-to-late 1980s, it appeared that Quincy would survive as a vital part of Cooper Industries' manufacturing mix, but that Gardner-Denver would disappear as a recognized name in American industry. Incredibly, both Quincy and Gardner-Denver would survive to write an exciting new chapter in global manufacturing in the 1990s and the twenty-first century.

Rationalization, Realignment, and Revitalization

The stage for the 1993 spin-off of Cooper Industries' Industrial Machinery Division was set with Robert Cizik's acquisition strategy in the late 1980s. In 1989, Cooper went into the debt markets and issued preferred stock to make two more major acquisitions: Champion Spark Plugs and Cameron Iron Works, a Houston-based rival in the oil fields. All told, Cizik raised $1.5 billion to buy the two companies.

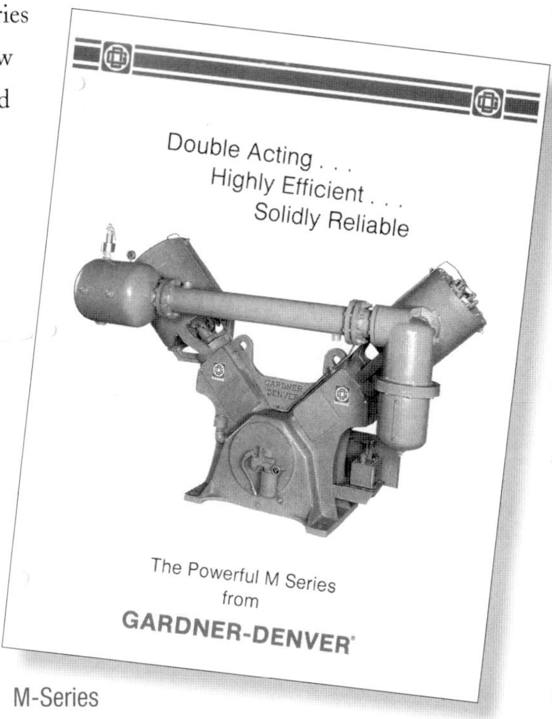

M-Series two-stage reciprocating compressor was introduced in 1986.

Cooper Industries decentralized its new acquisitions, reduced employment, and cut costs. By 1992, Cizik amassed a $1 billion war chest for another major acquisition. But another recession in 1991 and falling prices in the oil-patch caused a sudden sharp drop in Cooper's stock in the spring of 1992. Cizik, whose philosophy had always espoused buying diversified manufacturing operations, wringing out all of the excess costs possible and making capital investments to increase quality and productivity, was never shy about disposing of operations when he felt that they could no longer contribute to Cooper's bottom line.

As a result, Cooper began a year-long strategic planning effort to determine the company's direction in the 1990s.

Internally, the effort was dubbed "rationalization, realignment, and revitalization."

The changes were sweeping. When the plan was announced in October 1993, the actions affected literally all of Cooper's business segments. The seven-step program was designed to reduce costs, concentrate on areas of market strength, and address the competitive pressures of the 1990s.

Cooper kicked off the effort by divesting its Belden Wire and Cable business earlier in 1992. Late in the summer, it agreed to sell the Cameron Forged Products business. It announced a reevaluation of its McGraw Edison electric transformer operations. Cooper also implemented a number of cost-saving actions, including a host of plant closings and facilities consolidations affecting other business units. And Cooper announced it intended to engineer the tax-free spin-off of its Gardner-Denver Industrial Machinery Division.

After nearly twenty years in Texas and fourteen years as a subsidiary of Cooper Industries, Gardner-Denver was returning to its hometown of Quincy, Illinois, as an independent company.

The Quincy Herald-Whig, Wednesday, October 13, 1993 □ 4D

City/County

Officials praise Cooper decision to spin off G-D

By Tom Saul
Herald-Whig Staff Writer

Cooper Industries' decision to spin off its Gardner-Denver Machinery Division is good news for Quincy, say public officials and a financial analyst.

"They are a major employer and they provide excellent head-of-household jobs," said Mayor Chuck Scholz. "One of the primary goals of economic development is retention. It's easier to work with someone who is here in Quincy instead of being 1,000 miles away."

Houston-based Cooper Tuesday announced plans to spin off Gardner-Denver, which it acquired in 1979.

Mike Brennan, assistant branch manager and financial consultant for Smith Barney Shearson in Quincy, said the spin off will let Gardner-Denver Machinery Inc. react more quickly to changes in its market and among its customers.

Instead of having to generate profits for current parent Cooper, Gardner-Denver can concentrate on generating on-going business for itself and focus on its own needs, Brennan said.

"That's a very strong positive for Gardner-Denver," said Brennan.

"There is a misconception that this will be a small company. That's not true. It has annual sales of $160 million. That puts it well into the list of Fortune 1,000 companies. Now, it will be a Midwest company based in Quincy instead of a southern company based in Houston, Texas."

Brennan said the company's future will be tied with Quincy because the firm's chief executives and decision makers will live here.

"Their lives and our's will be intertwined," Brennan said. "They will be more attuned to what is happening here."

Location of its headquarters in Quincy "makes a difference in the company's attitude toward the community," added Harry Button, Quincy Area Chamber of Commerce president.

Scholz recalled city officials' concerns about Gardner-Denver during the summer's flooding. The riverfront plant was in danger of being overwhelmed by floodwater and the city sent about 100,000 sandbags and crews to throw them to save the facility, Scholz said.

"It's easier to work with someone who is here in Quincy instead of being 1,000 miles away."
— Mayor Chuck Scholz

"During the flooding we were afraid that if something did happen to them a decision would be made in some distant corporate office," Scholz said.

Great River Economic Development Foundation Executive Director Jim Mentesti said Gardner-Denver employees will now have a direct role in the future of the company.

Jim Nauert of Machinists Local 822, which represents workers at the plant, isn't sure what effect the spin off will have. Union members called him Tuesday "and I didn't have anything to tell them either," he said.

"They're concerned about what's going to happen," Nauert said.

Under terms of the spin off, shares of the new company will be distributed proportionately to holders of Cooper stock, said Cooper spokesman Ellen Orsburn. The spin off will take about six months to complete.

Cooper plans to buy up $300 million worth of its own outstanding shares. Current outstanding shares are valued at about $5.4 billion.

Ross Centanni, Gardner-Denver vice president and general manager, said the spin off could mean more jobs in the future.

Improvements in its compressor business have meant the addition of almost 50 jobs in recent months, said Director of Employee Relations Steve Krivacek. Thirty-seven jobs were added at its foundry in La Grange, Mo., and about 10 were added in Quincy.

Newspaper article in the *Quincy Herald-Whig*

The Industrial Machinery Division

January 6, 2009

When Cooper Industries acquired Gardner-Denver, it immediately expanded its organizational structure from three operations groups to six, including the Energy Services Group, the Mining and Construction Group, the Petroleum and Exploration Equipment Group, the Industrial Equipment Group, the Hand Tool Group, and the Airmotive Group.

In conjunction with the new organizational structure, Gardner-Denver's Industrial Machinery Division became part of Cooper's largest unit, the Energy Services Group. Headquartered in Mount Vernon, Ohio, Cooper's Energy Services Group produced centrifugal compressors, gas turbines, reciprocating pumps, and diesel and gas engines made and marketed through an international network of twelve manufacturing plants and some fifty sales and service centers located in fifteen different countries.

The only Gardner-Denver division to be incorporated into an already existing Cooper group, the Industrial Machinery Division was slated to continue to produce pumps, compressors, air blowers, and motors at plants in Quincy, Illinois, and York, Pennsylvania, as well as the LaGrange (Missouri) Foundry. Gardner-Denver would continue to serve mining, construction, industrial, and general exploration markets while the existing Cooper Energy Services Group plants would concentrate on the natural gas and oil pipeline transmission markets. By 1980, Cooper had installed more pipeline compression equipment in Europe and North America than any other company in the world.

Gardner-Denver's Drilling Machinery Division, along with its international components, was folded into Cooper's Mining and Construction Group, headquartered in Dallas. The Drilling Machinery Division included Gardner-Denver's plants in Denver, Colorado; Dallas, Texas; and Salem, Virginia; and produced seismographic and geophysical drilling equipment for the oil and mining industries. The International Division manufactured construction equipment, drill rods, and drill bits at plants in Johannesburg, South Africa; Rio de Janeiro, Brazil; and Mexico City, Mexico.

Cooper's Petroleum and Exploration Equipment Group also was primarily made up of Gardner-Denver divisions, including the company's Rotary Machine Division, Martin Decker Company and Demco Inc. The Rotary Machine Division in Dallas was assigned to continue manufacturing rotary blasthole drilling equipment, packaged pump and compressor units, and petroleum drilling equipment. Martin Decker would continue to manufacture torque and flow reading equipment for the petroleum industry at its Santa Ana, California, plant, while Demco Inc. would continue manufacturing valves and waste treatment equipment at its Oklahoma City facility.

Gardner-Denver's Pneutronics Division in Grand Haven, Michigan, and the Apex Machine and Tool Company in Dayton, Ohio, joined to form Cooper's Industrial Equipment Group. The Pneutronics Division was tasked with producing portable air-powered tools, grinders, air hoists, piston motors, and finishing tools at facilities in Grand Haven and Reed City, Michigan; Hicksville, Ohio; and Green Bay, Wisconsin. Apex Machine and Tool manufactured screwdriver bits and sockets for air and electric tools, as well as universal joints for machinery and aircraft applications at its factory in Dayton, Ohio.

Cooper's last two groups, Cooper Airmotive and the Cooper Group, operated completely independent from any of the Gardner-Denver Divisions. Cooper Airmotive, headquartered at Love Field in Dallas, Texas, repaired aircraft engines for a number of major commercial airline fleets, maintained airplanes, and distributed aircraft parts, components, and engines to eighty-six major manufacturers worldwide.

The Cooper Group, perhaps the company's best known unit, manufactured a wide variety of hand tools under a number of different national brands, including Boker Knives, Crescent, Weller, Wiss, Lufkin, Nicholson, and Xcelite.

The Cooper reorganization into six operating groups made a great deal of sense, and Gardner-Denver employees in the Industrial Machinery, Mining and Construction, Petroleum and Exploration Equipment, and Industrial Equipment groups were assured they would see little change in the day-to-day operations of their units. Cooper Industries had a reputation for decentralizing operations and allowing a fair amount of autonomy to local managers.

CHAPTER 8

Transformation

The Gardner Denver that emerged as an independent, stand-alone company in April 1994 was a shadow of the company that Cooper Industries acquired just fourteen years earlier. With annual revenues of just over $160 million, Wall Street's expectation was that shareholders would be rewarded within a year or two when a larger manufacturing firm acquired the company at a premium to the initial public offering.

Gone was the $650 million in annual revenues that doubled Cooper's size and allowed the Houston company to embark on its strategy of acquisitions in the 1980s. Gone was the Denver half of Gardner Denver, the rock drill business that had allowed the firm to diversify into profitable mining and construction segments of the economy. Gone were literally all of the acquisitions that Gardner Denver had made from 1955 to 1979.

For Cooper, the spin-off of Gardner Denver was simply a minor tweaking of the company's business strategy for the 1990s. "The slow growth of world economies and the steady increase of competition continue to challenge all of our businesses," explained Robert Cizik, Cooper's president and CEO, "despite ongoing productivity investments and cost improvements. These changes will lower operating costs and improve our earnings potential."

Cizik went on to note that the benefits of Cooper's mass cost reductions would more than offset the share earnings reduction from the divestiture of Belden and the spin-off of Gardner Denver. Cooper's CEO also explained that Gardner Denver no longer fit Cooper's mix of product lines.

"The Industrial Machinery Division we operate today is not the business we acquired in 1979," he said. "The division has been successful in refocusing operations as its primary business, equipment for oil and gas exploration and production, dwindled in the 1980s. Today, it primarily manufactures industrial equipment and is no longer the complement to our other oil field operations that it was intended to be."

While with Cooper Industries, Gardner Denver added Joy compressors, along with Sutorbilt and Duroflow blowers.

What Cizik didn't mention in the release was that Cooper Industries had originally intended the Industrial Machinery Division to be more than a petroleum equipment manufacturer. Cooper Industries had combined Gardner Denver's Air Compressor Division and Petroleum Equipment Division in 1985 to form the Industrial Machinery Division. Cooper added the OPI pump product line to the division that same year. Two years later, Cooper acquired the Sutorbilt and DuroFlow blower product lines and the Joy industrial compressor product line and added them to the Industrial Machinery Division.

Cizik did point out that the division was profitable and that it had the potential to regain even more market share in the future. "We believe that, as an independent company, Gardner Denver Machinery Inc. will be a viable entity in which management focus can be concentrated," he said, "and the benefits of a future turnaround in the oil and gas exploration and production equipment market can be fully realized."

The initial public offering was made in November 1993, and the new company—with a proud heritage stretching back almost 135 years—was incorporated as Gardner Denver Machinery Inc. The new firm's assets totaled about $175 million, and the company was headquartered in Quincy, the site of its major manufacturing facility.

The new company had strengths that were perhaps not immediately apparent to the corporate planners and accountants in Houston. The work ethic and pride of Gardner Denver's 1,100 employees wasn't to be underestimated. Ross J. Centanni, the Cooper Industries manager who was named the Quincy company's president and chief executive officer, assured employees that they would see very little change in day-to-day operations and management.

Ross J. Centanni, president and chief executive officer, after the spin-off from Cooper Industries

But Centanni did foresee the potential for growth. "I see this change as a significant opportunity for all of us at Gardner Denver Industrial Machinery," he told employees and shareholders. "I am convinced that we can reach our goal of becoming a total quality manufacturer, to provide total customer satisfaction, as a stand-alone company."

But first, Gardner Denver had to finish recovering from one of the worst natural disasters in the company's long history.

The 1993 Flood

While the planning and investor relations departments in Houston were putting the finishing touches on the Cooper Industries revitalization plan in the summer of 1993 that would include spinning off Gardner Denver, employees of the soon-to-be stand-alone company were desperately trying to save their workplace and homes.

It rained and snowed almost continuously over the Upper Midwest between the autumn of 1992 and the early summer of 1993. Precipitation across the Missouri and Upper Mississippi river basins was four to eight times above normal from the end of August 1992 to the summer of 1993. Already by April 1993, the Mississippi River was as much

Gardner Denver manufactures both reciprocating and rotary screw compressors that are used on locomotives for air brakes.

as six to ten feet above flood stage for much of its length between St. Paul and St. Louis.

Heavy rains throughout the late spring and early summer inundated five hundred miles of the big river above St. Louis. Even heavier rains in Iowa swelled all of the feeder streams and tributaries that flowed into the Mississippi. By the end of June, the Mississippi was closed to all commercial barge traffic between St. Paul and St. Louis, the first time that had happened since the record floods of 1965.

As the crest rolled inexorably downstream, Quincy and its neighbor community of Hannibal, Missouri, braced for the impact. By July 4, the Mark Twain Memorial Bridge carrying Interstate 72 over the Mississippi to Hannibal was closed by high waters. Quincy's time of trouble came a little more than a week later when the city's new Memorial Bridge was forced to close as the water lapped inches away from flooding the city's riverfront businesses. Gardner Denver workers feverishly placed sandbags around the more vulnerable parts of the riverfront plant, and as water circled around to cover Gardner Expressway, the plant was an island in a sea of dirty brown floodwater.

Ross Centanni explained to NBC News reporter Dawn Fratangelo that if floodwaters closed the city's second bridge, "we would probably lose 10 to 15 percent of our employees because they live in Missouri. We would have problems shipping product."

The great fear was the strength of the levees holding back the Mississippi. As the water continued to rise, cautious optimism reigned. The levees were holding. And then a levee at West Quincy, Missouri, broke and inundated more than fourteen thousand acres of prime Missouri and Illinois farmland. Tragically, the levee was sabotaged by a Missouri resident who told police he wanted to strand his wife in Missouri while he partied in Illinois. His night of partying earned him a life sentence in the Missouri Correctional Center in Jefferson City.

The breaking of the levee effectively isolated Quincy and Gardner Denver from the rest of the world. Employees who lived in Missouri undertook Herculean efforts to get to work. The twenty-minute commute from nearby Missouri communities to Quincy took three hours and more, since employees had to travel as far south as St. Louis or as far north as Iowa to find a bridge across the Mississippi River that was open.

Employees Sandra VanDyck and Peggy Durst, both Missouri residents, were typical of the dedication that Gardner Denver employees showed during the month-long flood event. Durst stayed with her entire family in an Illinois hotel room for several weeks on end so she could make it into work. VanDyck managed to catch a ride on a helicopter from Hannibal on the Monday of the flood. She remained in Quincy for the entire week, working twelve-hour days, until catching a ride back to Missouri on Friday.

In the end, the water did go down, and the last link west to Missouri was restored. Gardner Denver employees continued to sandbag the Gardner Expressway plant until the end of July. By then, most of the five hundred employees who had been working to shore up the levee protecting the plant were back at work in production. Still, about ten employees each day were assigned to levee patrol.

"Our employees have just been great through this thing and have made a lot of sacrifices," Steve Krivacek of Gardner Denver told reporters. "It's through their efforts that we're still in business."

Planning for the Future

Because of his background in strategic planning, Centanni had a clear vision of what he wanted Gardner Denver to accomplish in 1993 and after. First and foremost, Centanni wanted the new company to focus on quality.

Unrelenting global competition among compressor and oil equipment manufacturers dictated that Gardner Denver would be customer-driven. Even before the spin-off, Centanni told the Quincy Area Chamber of Commerce's Focus on Excellence series that customers "want quality. Not only do they want it, they demand it and expect it, and if you don't provide it to them they're going to go elsewhere. Our Total Quality Management movement is based upon survival. We feel that, in order to be here, we're going to have to become a total quality manufacturer."

Gardner Denver continued the enhanced training programs for employees that had been started by Cooper Industries in the early 1990s and began certifying vendors who supplied materials and parts to ensure quality. And Centanni sent a signal that the company would rely even more on employees for input and recommendations about making the TQM program more efficient.

"A key is getting your employees involved," Centanni said. "You saw that assembly line Henry Ford had. You also may know of Frederick Taylor, who was the father of industrial engineering. Those guys probably did more to harm U.S. industry than a lot of others. Some of you may think that's blasphemy, but what they did was they told the workers when you come in the shop, hang up your brain with your coat."

Centanni's philosophy stressed worker involvement. He rejected the idea that "management knew more about the shop floor than their employees, and workers should do only what they were told. There was no room for their input or ideas. That doesn't work in today's world."

Centanni's prescription for success in the cutthroat global world of compressor competition was that Gardner Denver would be a leader and not a follower. "We have a vision for Gardner Denver to be the lead dog in our industry," he told anyone who would listen.

To do that, Gardner Denver would have to get considerably larger—fast.

Growth by Acquisition

Much like Cooper Industries a decade before, Gardner Denver almost immediately began investigating potential acquisition targets. In 1994 and 1995, the company completed its first strategic plan, calling for a concentrated effort to grow through targeted acquisitions in the compressor and oil field equipment markets. Revenues at the time were still well below $200 million a year, and the company spent much of its first two full years as a stand-alone company getting its financial house in order.

In 1995, Gardner Denver arranged for a long-term loan to cover the short-term debt assumed in the asset transfer from Cooper Industries. The company put together a lending syndicate of five banks to finance any expansion efforts and worked with those banks when it made a series of year-end adjustments in 1994.

Gardner Denver began to expand by acquisition. Early in 1996, it made the first of twenty acquisitions that would transform the company from a Quincy-based manufacturer of compressors and pumps to one of the world's largest manufacturers of fluid management systems used in a wide variety of industries and services.

The company's first acquisition in January 1996 was Noramptco Inc., a Syracuse, New York, manufacturer of centrifugal blowers and exhausters. Gardner Denver's 1995 revenue was $192 million, and the sale of the LaGrange Foundry had provided the cash to leverage purchase of Noramptco and its primary operating subsidiary, the Lamson Corporation. At the time, Noramptco had annual revenues of $41.3 million, just less than a quarter of Gardner Denver's revenues.

Lamson's lines of blowers and exhausters were used primarily in industrial and wastewater application and complemented Gardner Denver's product lines by allowing the parent company to participate in the centrifugal segment of the air and gas handling industry, as well as in niche markets requiring air and gas management.

The ink was barely dry on the Noramptco purchase when Gardner Denver announced the completion of its acquisition of TCM Investments Inc. in a cash-for-stock transaction. Although considerably smaller than the Noramptco acquisition, the $7.2 million purchase of TCM, a Tulsa, Oklahoma, manufacturer and packager of pumps used in oil and gas well stimulation and cementing, was important because it gave Gardner Denver a physical presence in the U.S. oil field market. Ross Centanni, Gardner Denver's president and CEO, noted that the company had previously identified TCM as a future growth opportunity. "Not only will this acquisition provide a physical presence in the oil field market," Centanni told the financial press, "it will also allow Gardner Denver to become a sole source supplier of repair parts and remanufacturing services to some of the company's customers."

Overseas Expansion

With the first two acquisition targets safely in the fold, Gardner Denver began searching for an overseas expansion candidate that would give it a presence in global markets. Not since the acquisition of Padley & Venables and Bedford Steel nearly twenty years before had Gardner Denver enjoyed an international presence. And Centanni and the executive team were convinced that the real growth of Gardner Denver would come with the acquisition of overseas companies.

The strategy began to take place early in 1997 when Gardner Denver acquired Oy Tamrotor AB, a Finnish manufacturer of lubricated rotary screw compressor air ends and packages, located in Tampere, Finland. Once again, Gardner Denver financed the purchase price of 135 million Finnish marks—about $26 million—with cash.

Centanni noted that "the addition of Tamrotor provides Gardner Denver with a manufacturing base in Europe as well as market penetration in several European compressor markets. This transaction is reflective of the company's strategy of growing the business through synergistic product line acquisitions and penetrating international markets."

Left: Lamson blower

Right: Drilling pump package

Although Tamrotor was the only acquisition Gardner Denver closed in 1997, the Finnish company was a milestone for the growth through acquisition strategy. Revenues had increased to $218 million in 1996, a nearly 35 percent increase from 1994, and revenues of $135.5 million through the first six months of 1997 were indicative of the growth that the Lamson and TCM acquisitions were providing. Even more important, the new acquisitions were driving Gardner Denver net income growth. By the second quarter of 1997, earnings per share had increased 81 percent over 1996.

1998

What worked in 1996 and 1997 continued to work in 1998. Gardner Denver made two major acquisitions early in 1998 that helped strengthen the company's product line and bottom line.

The first was a definitive agreement for the purchase of the assets of Champion Pneumatic Machinery Company Inc., a subsidiary of CRL Industries Inc. Located in Princeton, Illinois, about halfway between Joliet and the Quad Cities, Champion was a leading manufacturer of lubricated and non-lubricated single-acting, single- and two-

50000
GARDNER-DENVER

Engineered package

stage compressors. Champion products served the industrial markets, service support industries, oil-free applications, and consumer markets. For the first time, the $24 million purchase price was subject to a federal antitrust ruling and was treated as a cash purchase for accounting purposes.

The addition of Champion provided Gardner Denver with a more complete product offering to penetrate the low-horsepower compressor market. More importantly, the Champion acquisition opened new markets and channels of distribution to Gardner Denver, mainly the automotive aftermarket and catalog sales. As Gardner Denver continued its acquisition strategy through the final years of the twentieth century, the company was always alert to opening new distribution channels for its products.

Champion compressors are used in the automotive service industry as well as the construction and medical industries.

At approximately the same time as the Champion acquisition, Gardner Denver reported that it had acquired substantially all of the assets of Geological Equipment Corporation for $12 million. Located in Fort Worth, Texas, Geoquip, as it was known to customers, was a leading manufacturer of pumps, ranging in size from 250- to 2,400-horsepower. Primary customers of the Texas firm were well service and water blast industries. Geoquip also remanufactured pumps and provided repair services to its customer base.

The next major acquisition was again overseas, in March 1998. Gardner Denver purchased the Wittig Division of Mannesmann Demag AG in Schopfheim, Germany. Wittig, which manufactured rotary sliding vane compressors and vacuum pumps to serve the truck blower market for liquid and dry bulk conveyance, was one of the older companies in the pump and compressor manufacturing business. Established in 1885, Karl Wittig GmbH earned a grant of patent of its sliding vane compressor principle in 1908 and began licensing its manufacturing system to foreign countries in 1913. In 1956, Mannesmann AG acquired Wittig, and the company produced its first rotary vane compressors with plastic vanes and oil injection cooling in 1964.

New Markets, New Opportunities

Acquisitions continued into 1999. In April, Gardner Denver acquired Allen-Stuart Equipment Company Inc. Located in Houston, Allen-Stuart designed, fabricated, and serviced custom-engineered packages for blower and compressor equipment in air and gas applications. The addition of Allen-Stuart gave Gardner Denver the added capability of supplying its customers with engineered packages.

Later in April 1999, Gardner Denver acquired a second Houston company, Butterworth Jetting Systems Inc. Butterworth was a respected manufacturer of water-jetting pumps and systems serving the industrial cleaning and maintenance market. The cash-for-stock transaction for the $12 million in annual Butterworth revenues gave Gardner Denver an entry into the rapidly growing water-jetting business.

Applications in the market included runway and ship-hull cleaning, concrete demolition, and metal surface preparation. The high rate of growth in the industry was driven by a number of factors, including the environmental advantages of water blasting compared to sandblasting, the ease of cleanup, and the inherent effectiveness of water jets.

Ross Centanni noted that the Butterworth acquisition enabled Gardner Denver to expand its position "in a rapidly growing segment of the industrial market, thereby reducing the company's dependence on the petroleum business."

The final acquisition of the twentieth century came in October 1999 when Gardner Denver announced it was acquiring the assets of Air Relief Inc., a Mayfield, Kentucky, provider of replacement parts and service for centrifugal compressors.

The Kentucky firm had $11 million in annual income and was designed to "speed Gardner Denver's penetration of the centrifugal compressor market, which we will enter through our recently executed long-term strategic alliance with Samsung Aerospace Industrial Company of South Korea," the company told the financial media. "This alliance provides Gardner Denver with technology and manufacturing rights of the Samsung centrifugal compressor for North and South America and Western European countries. The addition of centrifugal compressors is a natural extension of our existing reciprocating and rotary screw compressor products lines. Air Relief will add key centrifugal compressor engineering, assembly, sales, and service capabilities to enable us to better support this new product line. We are excited about the opportunities that the acquisition of Air Relief and alliance with Samsung present, and look forward to improved compressed air solutions and services offered for our customers."

The Air Relief acquisition closed out a remarkable five-year period for Gardner Denver in which the Quincy-based manufacturer of pumps and compressors acquired nine companies in the United States and Europe, integrated them into the parent company and increased its presence in a number of markets the company had never served before.

The results were little short of spectacular. In 1999, Gardner Denver's revenues were $324 million, double the $160 million in annual revenues the company had carried into the IPO back in 1994. The company's five-year track record of revenue increase demonstrated the success of the Gardner Denver growth strategy, and the company's 1999 net income of $18 million proved the company's ability to focus on long-term growth in earnings and equity.

Gardner Denver's network of highly skilled service technicians service our products worldwide.

"As we integrate these acquisitions," the company told shareholders early in 2000, "we continue to leverage our manufacturing skills and strong brand names to generate higher revenues. Moving into the next millennium, we will continue to enhance our business growth through additional strategic acquisitions."

If Gardner Denver's growth had been impressive during the first five years of the company's six-part strategic plan, growth during the next five years would be phenomenal.

The First Executive Team

The 1994 executive team: left to right, Dennis Shull, Ross Centanni, Helen Cornell, and Harry Jefferson on factory floor. *Courtesy of Maxwell & Bellis Photography*

With the 1993 flood cleanup behind them, the employees of Gardner Denver embarked upon essentially building a new company from scratch. For all intents and purposes, the executive team of the new company was the same group of managers who had guided the fortunes of the Industrial Machinery Division during the late 1980s and early 1990s.

Ross J. Centanni was a thirteen-year veteran of Cooper Industries. A graduate of Louisiana State University with a BS degree in industrial technology and an MBA, Centanni joined Cooper Industries in 1980 as director of corporate planning. He had spent two years with Occidental Petroleum in Houston and seven years with B.F. Goodrich in Akron, Ohio, before joining Cooper.

Centanni worked with Cooper's executive team during the early 1980s as the company put together the strategic acquisition plan that created the diversified manufacturing behemoth of the early 1990s, so he was no stranger to the ins and outs of mergers and acquisitions. In 1985, he was named director of marketing of the newly created Industrial Machinery Division. Five years later, Centanni was elected vice president and general manager of the division.

Centanni's successor as director of marketing in Cooper Industries' Industrial Machinery Division joined Gardner Denver in 1975 prior to the acquisition by Cooper. J. Dennis Shull was a graduate of Northeast Missouri State University with a BS degree. He also earned an MA degree in business from Webster University in St. Louis.

Harry Jefferson III, a longtime Gardner Denver employee, served the company as vice president of finance and treasurer when the company was spun out as a stand-alone firm in late 1993. Jefferson, a certified public accountant, earned a BBA degree in industrial management from the University of Texas, an MBA in finance and accounting from Southern Methodist University, and an MA in business from Webster University. Jefferson was the director of finance for Gardner Denver's Industrial Finance Division from 1985 to 1993.

High-pressure water is used for hundreds of different industrial cleaning applications.

CHAPTER 9

Gardner Denver on the Global Stage

In August 2006, Ross Centanni, the president, CEO, and chairman of Gardner Denver, sat down with a reporter from the prestigious *Wall Street Transcript*. Centanni had just completed his thirteenth year at the helm of Gardner Denver, a position he had held since the Quincy-based manufacturer of pumps and compressors had been spun off to the shareholders of Cooper Industries Inc.

Those thirteen years had been among the most productive in the company's nearly century-and-a-half of existence. When Cooper Industries cut loose what essentially was a shadow of the company it had bought in 1979, Gardner Denver was a small manufacturer of reciprocating and rotary compressors and blowers for various industrial applications and pumps for industry and oil field applications with a headquarters office in Quincy; manufacturing plants in Quincy and Sedalia, Missouri; a remanufacturing facility in Fishers, Indiana; and warehouses in Memphis, Tennessee; Oklahoma City, and Singapore. The company had annual sales of $160 million in 1993 and seemed to be a candidate for a quick sale to another, larger industrial manufacturer.

Instead, Centanni and the Gardner Denver executive team had engineered a growth-by-acquisition strategy that had allowed the company to grow significantly in little more than a decade. From 1996 to 2006, Gardner Denver had acquired twenty companies, integrated them into the parent company structure, and lowered costs to make them a part of one of the most competitive manufacturers in the world.

Indeed, by 2006, Gardner Denver revenues had climbed to $1.67 billion, a 37 percent increase from the $1.21 billion reported in 2005 and a sevenfold-plus increase from the $218 million the company reported in 1996, the beginning of its remarkable record of earnings and profit growth. Net income of $133 million was nearly double the $67 million reported in 2005 and more than eight times the $17 million the company had earned in 1996.

Gardner Denver returned to the New York Stock Exchange on August 14, 1997. August 14 was also the day that Robert Gardner got the company's very first patent for his flyball governor in 1860.

Thomas pumps help make oxygen concentrators more portable, improving the quality of life for those who need one.

Even more impressive, Gardner Denver had fueled its growth with a tremendous amount of diversity, both in terms of the industries it served and the geographic location of its customers.

In 2006, the energy sector was still important, accounting for 23 percent of Gardner Denver's revenues. That percentage was matched by industrial manufacturing, while transportation end-users accounted for 14 percent of revenues. The remaining 40 percent of Gardner Denver's revenues were derived from a host of end-users, including medical, food and beverage, environmental, and automotive services. None contributed more than 7 percent of the company's revenues.

The company described itself as a "leading global provider of high-quality industrial equipment, technologies and services to a broad and diverse customer base through a family of highly recognized brands and strong distribution networks."

The global characterization was apt. In 2006, less than half the company's revenues came from end-users in the United States. Customers in Europe accounted for more than 40 percent of Gardner Denver's revenues, and Asia was the fastest-growing revenue producer for the company.

Gardner Denver equipment could be found all over the world. Workers on the mammoth Three Gorges Dam in China's Yangtze River Valley used Gardner Denver fluid transfer products. Passengers on the high-speed trains that crisscrossed the European Continent relied on Gardner Denver oil-lubricated vane compressors for safety and efficiency. Patients needing oxygen therapy were allowed a new level of freedom with Inogen's portable oxygen concentrator, made possible by a 1.8-pound compressor developed by Gardner Denver Thomas Products Division.

Gardner Denver's variable-volume multistage centrifugal blowers helped power wastewater aeration systems in Poland and the Middle East. Gardner Denver rotary screw compressors were frequently found in the world's growing commercial dry bulk carrier fleet.

In Lower Manhattan, the rescue and recovery crews working in the rubble of the World Trade Center following the September 11, 2001, attack utilized mammoth Gardner

Gardner Denver equipment was used by the rescue and recovery crews after the 9-11 attack.

Denver pumps and compressors to power their agonizing work.

Gardner Denver's emergence was little short of remarkable. And it had occurred because the company had formulated a relatively simple strategic plan and followed that plan faithfully.

That plan, which had been put into place in the mid-1990s shortly after Gardner Denver was spun out from Cooper Industries, outlined six mutually supportive strategies. As part of the plan, Gardner Denver vowed to acquire complementary products, pursue international markets, embrace new technologies, reduce costs, accelerate new product development, and manufacture proprietary products.

Worldwide Sourcing

By the early twenty-first century, Gardner Denver had made cost reduction a key element in its corporate strategic planning. The adoption of worldwide sourcing increased Gardner Denver's international competitiveness "through reduced raw material costs, greater diversity of component sources and stronger relationships in key markets," the company told shareholders in early 2004.

The company's global casting team, made up of corporate staff and key manufacturing representatives, scoured the world to source lower costs components from Eastern Europe, particularly Romania and the Czech Republic. The team also sourced castings from China, India, and Korea.

Embracing New Technology

Keeping costs low was only one thrust of the multipronged strategic initiatives that accounted for Gardner Denver's spectacular growth between 1996 and 2006. Another element of the strategic plan involved embracing new technology.

Centanni explained that Gardner Denver's technology was "really evolutionary, not revolutionary, but we don't want anyone to leapfrog us. Furthermore, we use new technologies to help achieve our other strategies, such as product development."

Gardner Denver's investment in analytical capabilities to increase the efficiency and speed of new product development helped enhance the company's responsiveness and competitiveness as it entered new market segments and developed new products to serve those market segments.

In 2004, Gardner Denver began acquiring several suites of analytical software that enabled the company's design engineers to predict such factors as bending, flexing, and shaft-loading, the behavior of fluid and gas under pressure and temperature fluctuations under a variety of inlet conditions and pressures. In 2006, the company began making use of software that allowed engineers to understand and reduce noise, vibrations, and harshness. The end result of these new technologies was an increased ability to model new products more fully during the design phase.

"Taken together," the company told shareholders, "these capabilities enable us to dramatically reduce development time and cost by evaluating computer-modeled versions of new products instead of a series of physical prototypes." The addition of a state-of-the-art rapid prototyping system further accelerated new product development by facilitating pattern making, assembly, and tooling for machining.

By early 2007, Gardner Denver was experiencing faster time to market, lower development costs, and far fewer warranty issues. When it launched its new line of variable speed compressors in 2006, the company brought seventeen models to market in thirty-two months, a dramatic reduction from the previous average time-to-market of two models per year.

Engineers employ stereolithography for new product development.

Sticking to Its Knitting

The heart and soul of Gardner Denver's strategy continued to be timely acquisitions of candidates that lend strength to the company's product mix and bottom line.

Between 1996 and 1999, Gardner Denver acquired ten companies. Each acquisition was well thought out. Each manufactured a complementary product. Each was synergistic with Gardner Denver's primary business.

"We have to understand the markets," Centanni explained. "We have to understand the product technology. We have to understand the customer base, and we won't stray from our basic markets. We stick to our knitting, so to speak."

The company would follow the targeted acquisition growth strategy to acquire ten more companies in the twenty-first century.

Total Worldwide Organic Growth

Gardner Denver's first acquisition in the new millennium was Invincible Airflow Systems. The company acquired the Baltic, Ohio–based manufacturer of single and fabricated multistage centrifugal blowers and engineered vacuum systems in January 2000. Invincible extended Gardner Denver's product offering for the industrial cleaning market and introduced the company's growing line of centrifugal blowers to new markets.

Gardner Denver followed up the Invincible acquisition with the acquisition of Jetting Systems & Accessories Inc. and CRS Power Flow Inc., both located in Houston, in April and July of 2000 respectively. Both Texas companies manufactured aftermarket products of the water jetting industry, and both complemented Gardner Denver's product offering for the water jetting market. In addition, the two acquisitions further leveraged Gardner Denver's commitment to being a full-service provider in the water jetting market segment.

The acquisition strategy forged ahead in 2001. On the home front, the company acquired Hoffman Air & Filtration Systems, a Syracuse, New York, provider of multistage

centrifugal blowers and vacuum systems serving industrial and wastewater treatment markets. Gardner Denver noted that Hoffman "added a highly recognized brand name" to the company's line of centrifugal blowers. The integration of Hoffman into existing Gardner Denver operations in Georgia and Ohio quickly created significant manufacturing and marketing synergies.

Centrifugal blowers are used in wastewater treatment facilities worldwide.

Gardner Denver is a leader in oil-free compressor technology for PET bottle blowmolding, providing more energy efficiency at lower service and support costs than competitive products.

Belliss & Morcom

A second 2001 acquisition allowed Gardner Denver to continue pursuing its strategic target of pursuing international opportunities. The company acquired Belliss & Morcom, Gloucester, England, to give it a foothold in the rapidly growing market for plastic bottle consumption in Great Britain and Europe. Belliss & Morcom manufactured and distributed reciprocating air compressors used for a variety of niche applications, including polyethylene terephthalate (PET) bottle blowing, breathing air equipment, and compressed natural gas.

Ross Centanni told shareholders, "Belliss & Morcom broadens our product offering, strengthens our distribution and service networks, and increases our participation in market segments that have the potential to grow faster than the overall industrial economy."

Belliss & Morcom sported an industrial tradition that stretched back as far as that of Gardner Denver. Founded by G. E. Belliss in Birmingham in 1852 to supply boilers and steam engines to the British Navy, the firm renamed itself Belliss & Morcom in 1884 when Alfred Morcom, a well-known British engineer, joined Belliss in management of the company. In 1890, Morcom and Chief Engineer Albert

C. Pain perfected the first forced-feed lubrication system, which revolutionized the performance and endurance of steam engines. For the next fifty years, Belliss & Morcom's improvements were found on steam engines serving the electric utility industry in Europe and the United States.

Belliss & Morcom began manufacturing air compressors in 1904 and, during much of the twentieth century, designed and built turbines of all sizes and types. Turbine compressors designed by the firm helped increase steam pressures and temperatures, as well as the rpm of turbines used in electric power generation and to power vessels engaged in maritime commerce.

By the 1960s and 1970s, Belliss & Morcom had established a reputation as one of the most innovative manufacturers of air compressors in the world. Its 1962 introduction of the "Twin V" model resulted in a significant reduction in weight in compressor applications around the world.

Gardner Denver made a second acquisition of a respected British company when it purchased Syltone LLC, a publicly held manufacturer of loading arms and transportation products. Formed in 1999 from the merger of Drum Industries and Emco Wheaton, Syltone helped Gardner Denver create a leading global provider of solutions for customers in the transportation and bulk product handling industries. Gardner Denver's sixteenth acquisition since 1996, Syltone also was the largest, with annual revenues of more than $150 million.

In Europe, the superiority of our oil-lubricated vane compressors, together with our solid inmarket presence, has enabled us to expand the applications for this product into braking system compressors for passenger railways and trams.

The acquisition also gave Gardner Denver access to two highly respected brand names. Drum Industries, founded in the United States in 1973, provided mobile power systems and specialty unloading equipment to a wide variety of customers in the transportation marketplace. Drum manufactured and distributed hydraulic drive equipment for mobile hydraulic power applications, as well as pumps, blowers, and compressors for the discharge of tank trucks used in the food, chemical, bulk feed, and cryogenic liquids trade.

Emco Wheaton was the world's largest manufacturer of loading and unloading arms. Founded in 1935, the company helped engineer applications for loading and unloading arms serving the truck, rail, and marine transport sectors of the market. Emco Wheaton's solutions to liquid and vapor handling issues were found throughout the petroleum,

chemical, and compressed gas industries in the United States and throughout the world.

In addition to the British acquisitions between 2001 and 2003, Gardner Denver beefed up its growing presence in Asia by opening a new logistics and warehouse facility on the outskirts of Shanghai, China, in 2004 to package compressors and blowers for a wide range of applications in Chinese industry.

"China is quickly becoming the world's factory floor," Centanni told shareholders early in 2004, "and this manufacturing boom requires equipment and support from American companies on the ground in China."

Leveraging a World of Opportunity

Gardner Denver continued to pursue global opportunities in the period from 2004 to 2006. Syltone's

Loading arm systems, including swivel joints, couplers, hydraulic drives, and electric controls from Gardner Denver make the loading and unloading of liquids and compressed gases from ships and ocean-going supertankers faster, safer, and cleaner.

record as Gardner Denver's largest acquisition was eclipsed in 2004 when the company acquired all of the outstanding ownership interest of Nash Elmo Holdings LLC from Boston-based Audax Private Equity for a purchase price of \$223.5 million. Formed through the 2002 combination of the operations of the Nash Engineering Company, Trumbull, Connecticut, and Elmo Vacuum Technology GmbH, Nash Elmo was the largest producer of liquid ring pumps in the world, as well as a leading manufacturer of side channel blowers.

Nash Elmo fit the Gardner Denver acquisition strategy in two regards. It manufactured a line of complementary products, and it had a global presence. The company operated major plants in Germany, China, and Brazil and was a major supplier of energy-efficient vacuum and blower systems to the Indian Oil Company. In China, the new Gardner Denver packaging and assembly facility in Shanghai, which opened about the time the company announced the Nash Elmo purchase, was an excellent complement to the existing Nash Elmo factory in the region. Together, the two facilities would be used to provide liquid ring pumps to the rapidly increasing Chinese power generation sector.

Compressor operation in Shanghai

Nash Engineering Company was founded in 1905 in Norwalk, Connecticut.

The creation of a sales presence in Shanghai in conjunction with the completion of the new assembly plant and the acquisition of Nash Elmo and its Chinese factory was consistent with Gardner Denver's strategy of building distribution channels. "Our competitive advantage is the distribution channels that we use," Centanni explained in his 2006 interview with the *Wall Street Transcript*. "We sell to many market segments, and each one requires a different sales channel."

That was amply demonstrated when Gardner Denver emerged in South Africa in 2004. The 2001 acquisition of Belliss & Morcom and its distribution facilities in South

Africa allowed Gardner Denver to establish contacts and expand markets beyond the PET plastic bottle industry. By 2004, Gardner Denver was selling breathing and industrial air applications to the country's mining industry.

Gardner Denver achieved a major breakthrough in diversification in the early spring of 2005 when it signed a definitive agreement to acquire Thomas Industries Inc. for $40 per share. Headquartered in Louisville, Kentucky, Thomas Industries, a New York Stock Exchange company, was a worldwide leader in the design, manufacturing, and marketing of precision-engineered pumps and compressors.

Thomas Industries gave Gardner Denver entry into a number of attractive end markets, including medical/laboratory, general industrial, printing, environmental,

Gardner Denver Nash liquid ring pumps help refineries capture and recycle flare gas.

and packaging sectors. The company's primary U.S. manufacturing facilities were located in Sheboygan, Wisconsin; Monroe, Louisiana; Skokie, Illinois; and Syracuse, New York. Thomas operated a quartet of factories in Germany and had recently completed the construction of a manufacturing facility in Wuxi, China, to serve the Asian market.

Thomas Industries allowed Gardner Denver to penetrate the fast-growing medical markets. Thomas compressors were used in many of the leading brands of stationary oxygen concentrators, and small Thomas pumps were commonly used in hospital patient monitors for gauging blood pressure, heart rate, and other applications. Hospital applications marketed under the Rietschle brand were used in vacuum pump operation for surgery suite applications, ventilators, and aspirators.

A second 2005 acquisition brought Bottarini S.p.A. into the Gardner Denver family. Bottarini, a leading packager of industrial air compressors located in Milan, Italy, provided Gardner Denver with expanded capacity and a low-cost compressor assembly operation to better serve the European market.

Gardner Denver used the period from 2006 to 2008 to consolidate the major acquisitions it had made since the beginning of the new century. It made several small acquisitions, including Todo, a manufacturer of self-sealing couplings in 2006, and Best Aire Inc., an Ohio-based distributor of compressed air and gas products. Perhaps the best example of the success of Gardner Denver's acquisition strategy came in February 2006 when the company's board of directors approved a two-for-one split of the company's common stock effective June 1.

Gaining Recognition

Gardner Denver's growth was quickly gaining recognition. For Ross Centanni, it was all sweet vindication of a more-than-decade-long strategy that Gardner Denver had adhered to religiously. "You face challenges every day, but, basically, you just have to focus on continuous incremental improvement. If you try to create too much change in a short period of time, you could hurt the organization. Therefore, we are very methodical when it comes to implementing change so that our business remains well controlled. If you're not under control, then you're in trouble."

If anything, Gardner Denver was a model of control as it approached its sesquicentennial celebration.

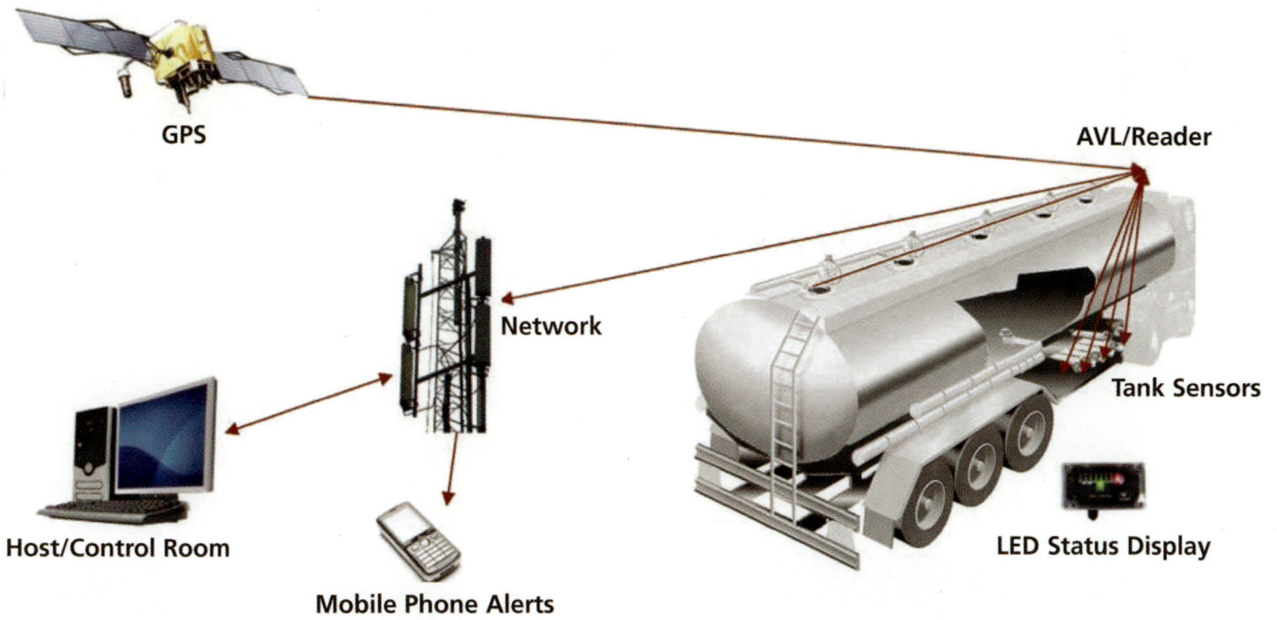

Emco Wheaton's wireless Sealed Cargo System provides remote monitoring of fuel tank valves, allowing companies to know when valves are opened and closed. It can also prevent access to the fuel when the truck is outside of a specific location or "geographical fence."

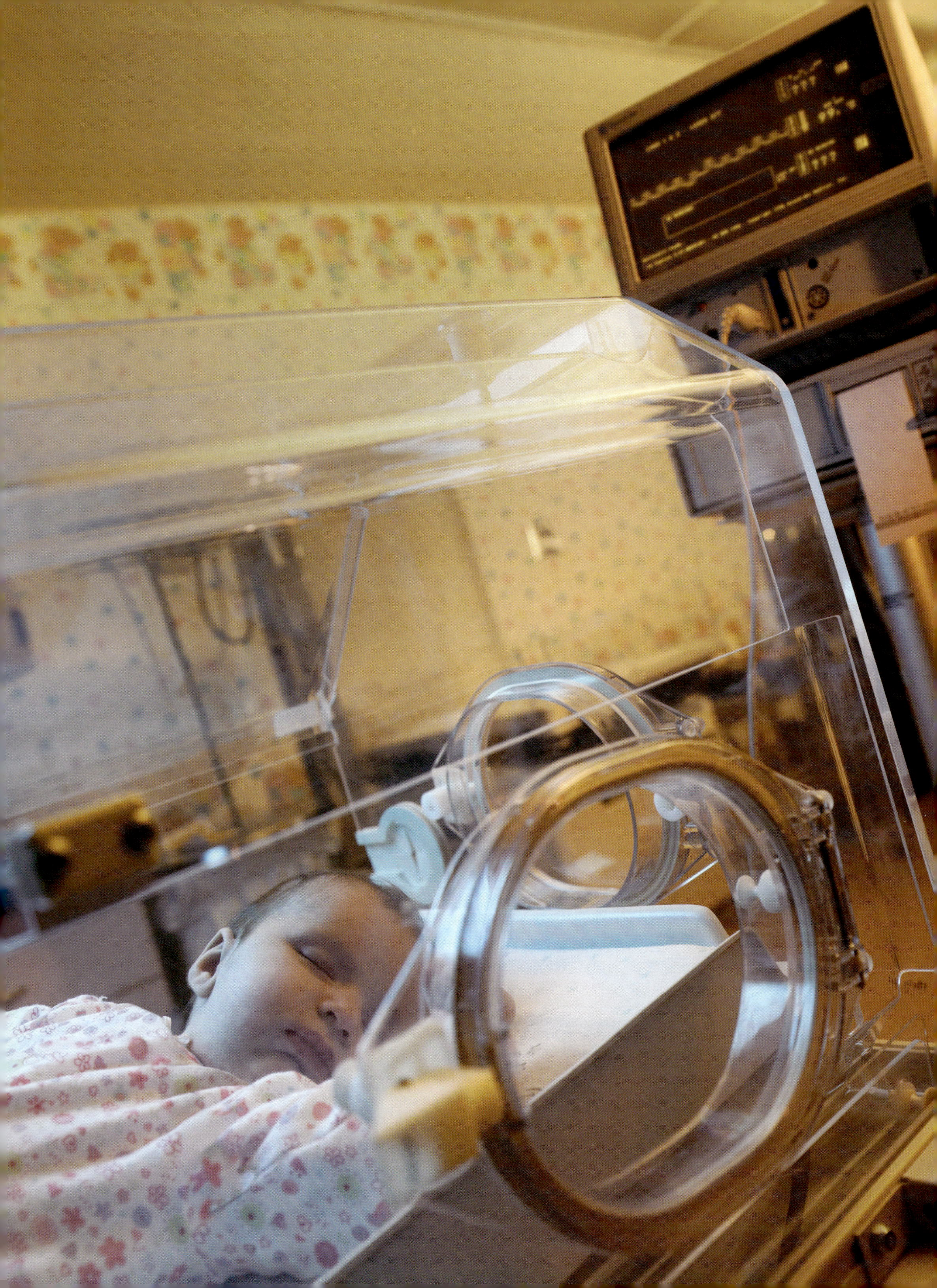

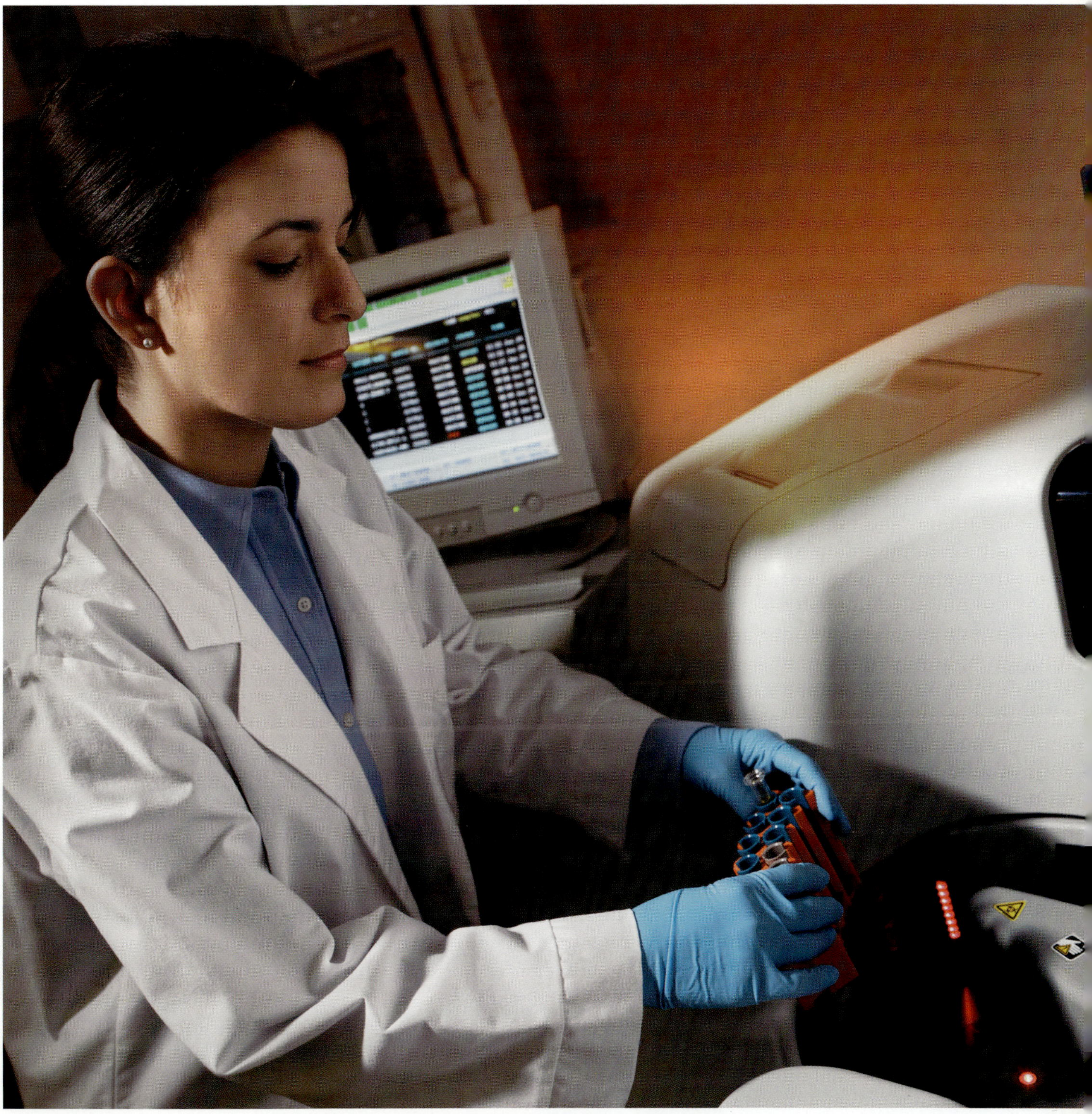

Gardner Denver works behind the scenes in the medical sector, providing key components for equipment used in patient care and monitoring in hospitals and at home, in laboratories, and in pharmaceutical research and production.

Our two-stage vacuum screw pumps produce more solar panels at a quicker pace, using less energy than the competition.

Elmo blowers have revolutionized water and energy management in breweries. When boiling below atmospheric pressure, the primary energy consumption of this process can be reduced by up to 70 percent and the carbon dioxide emission by up to 40 percent. This dramatically reduces energy requirements and costs, and also protects the environment.

Biodiesel is produced from renewable raw materials, mostly rapeseed oil, also known as canola oil, which is extracted from the plant's seeds through an oil press. In Germany, about eighteen hundred filling stations offer biodiesel. First, however, this fuel of the future must be loaded into tank vehicles for transport. Emco Wheaton manufactures and delivers custom-designed loading arms, fitted to the manufacturer's needs.

Gardner Denver's oil-free compressors help maintain the purity of drinking water as part of a critical water transmission system, which distributes desalinated water to residents and religious pilgrims in cities throughout Saudi Arabia.

Barry L. Pennypacker

When Gardner Denver investors received their 2007 annual report in the mail during the spring of 2008, they received confirmation of news from a January 2, 2008, press release when they read Ross Centanni's annual letter to shareholders.

"In closing," he said, "I would like to welcome Barry L. Pennypacker, who succeeded me as Gardner Denver's President and Chief Executive Officer in January 2008."

Centanni, who was named Gardner Denver's executive chairman, announced that he would transition the day-to-day operational responsibilities of Gardner Denver to Pennypacker and "begin to focus my attention on strategic growth opportunities for the Company, including potential acquisitions." In May, Centanni was named chairman emeritus, a position he held until his retirement on January 2, 2009.

Pennypacker joined Gardner Denver from Wabtec Corporation, where he had served as a group vice president for the global provider of technology-based equipment and services for the rail industry since 2002. Pennypacker had joined the former Westinghouse Air Brake Company in 1999 as a vice president of performance. Prior to his tenure at Wabtec, Pennypacker had served with Stanley Works from 1997 to 1999 and with Danaher Corporation from 1992 to 1997 in a number of senior management positions. Pennypacker earned a bachelor of science degree in operations management from Pennsylvania State University and an MBA in operations research from St. Joseph's University.

The new CEO quickly established the reputation of being a hands-on manager. During his first sixty days on the job, Pennypacker traveled to Gardner Denver locations around the world. "Out of the twelve hours I was at Nuremberg," he said, "I spent eight hours on the shop floor. During my three days at Shopfheim, I spent two days on the floor."

Pennypacker told *Gardner Denver World* shortly after arriving at Quincy that employees could "expect to see significant emphasis on improving operations."

President and CEO Barry Pennypacker prefers the shop floor to the Executive Board Room.

CHAPTER 10

Gardner Denver Today and Tomorrow

In December 2008, as it prepared to begin its 150th year of operation, Gardner Denver announced a series of restructuring initiatives designed to cope with the deteriorating global financial situation and the integration of the company's October 2008 acquisition of CompAir Holdings Ltd.

The CompAir acquisition was the latest in a string of nearly two dozen that Gardner Denver had engineered between 1996 and 2008. A leading global manufacturer of compressed air and gas solutions, CompAir served customers from its Redditch, United Kingdom, headquarters in a number of economic sectors through three business divisions: Industrial, Hydrovane, and Reavell.

The British firm's Industrial Division manufactured an extensive range of products, including oil-injected and oil-free rotary screw compressors, piston compressors, and portable rotary screw compressors used in oil and gas exploration, mining and construction, power plants, and general industrial applications.

The Hydrovane Division is a leading manufacturer of rotary vane compressors used in Original Equipment Manufacturer (OEM) applications such as snowmaking and mass transit, as well as industrial applications. The Reavell

Division is a leading manufacturer of high pressure reciprocating compressors producing air at delivery pressures in excess of 400 bar. Reavell's compressors are used in a number of applications, including compressed natural gas, industrial gases, breathing air, and in naval, marine, and defense market segments.

Gardner Denver paid $395 million to acquire CompAir in a combination of cash and the assumption of existing company debt. Gardner Denver President and CEO Barry L. Pennypacker noted that the British company was "an exciting acquisition for Gardner Denver. The company has a reputation for high-quality innovative products and best-in-class customer service, which strongly complements Gardner Denver's existing operations."

CompAir facility in Redditch, United Kingdom

CompAir serves a diversified global customer base of distributors, OEMs, end-users, and engineered system customers with a relationship strategy that focuses on product innovation and value-added designs. Its primary manufacturing facilities are located in Redditch and Ipswich, United Kingdom; Simmern, Germany; Ocala, Florida; and Shanghai, China.

Pennypacker noted that Gardner Denver believes "there is more opportunity to grow the profitability of this business post-transaction through lean enterprise techniques, low-cost country sourcing, and capturing integration synergies." Gardner Denver financed the acquisition through the use of excess available cash and new syndicated credit facilities.

By the time the CompAir acquisition closed in October, however, the world of business finance had been turned upside down, and companies with significant amounts of debt on their balance sheets were struggling to survive.

Restructuring

The global credit crunch that emerged from the wreckage of the collapse of Lehman Brothers, the U.S. investment bank, was unprecedented in its severity at least since the end of World War II. Almost overnight, investment grade bond markets worldwide contracted, and banks stopped lending—even to each other.

The U.S. Federal Reserve, the European Central Bank, and government banking authorities in China, Japan, and other Asian countries began injecting massive amounts of capital into the banking system, in an attempt to stimulate the flow of credit. Interest rates at government bank discount windows dropped to zero.

The subprime mortgage crisis in the United States and the lack of credit created an unprecedented contraction in the price of housing. Global automotive manufacturers watched as automobile sales in the United States dropped from 16 million units annually to 13.8 million units in 2008.

By year-end, the U.S. economy, and that of much of the developed world, was in a full-blown recession. Consumer spending plummeted, leaving much of the U.S. retail industry high and dry during the all-important holiday buying season. The price of oil, which had reached $140 a barrel during the summer of 2008, imploded as demand evaporated; by January 2009, oil was selling on world markets for $35 a barrel. Unemployment in the United States and the developed country skyrocketed as businesses found themselves strapped for cash.

Gardner Denver anticipated the credit crunch and resulting recession. The company began aggressive cost reduction activities in the third quarter of 2008 in recognition of a slowing global economy. Continued deterioration in end market conditions in excess of previous expectations caused the company to respond in the fourth quarter with further cost reduction actions, including a reduction of its global salaried work force. In addition, Gardner Denver implemented a hiring freeze and strict controls on discretionary spending.

As part of its integration of CompAir, the company closed a manufacturing plant in the United Kingdom.

High-pressure air compressors serve markets needing breathing air. Gardner Denver manufactures breathing air compressors used for scuba diving and for rescue efforts.

It notified approximately 9 percent of its global salaried workforce that their positions would be eliminated during the first nine months of 2009. Finally, Gardner Denver expected to record an additional charge of $5 million during the fourth quarter of 2008, primarily related to severance expenses associated with the reduction of the global salaried workforce.

For Gardner Denver, the financial upheaval of late 2008 was truly global. Unfavorable changes in foreign currency exchange rates, due primarily to the strengthening U.S. dollar relative to the euro and the British pound sterling, lowered equivalent earnings of Gardner Denver's foreign operations more than anticipated. As a result, the company restated its fourth quarter earnings estimates to reflect the lower expected numbers.

"Orders for our products serving industrial end market segments, particularly in the U.S. and Europe, have slowed significantly faster than was originally expected," said President and CEO Barry Pennypacker. "Demand for these products tends to correlate with the level of manufacturing capacity utilization. The rapid decline in industrial production in the U.S. and Europe has resulted in reduced levels of capacity utilization and deferred purchases of capital equipment, such as compressor packages. The decline in demand has translated into lower revenues, operating margin, and diluted earnings per share. At present, we are uncertain how long orders will remain at these depressed levels. However, management has proactively identified opportunities to reduce costs, and we plan to continue to rationalize our manufacturing footprint."

Pennypacker went on to note, "as a result of these efforts, we expect to be well positioned when demand for industrial equipment recovers and remain focused on progressing toward the company's long-term operating margin improvement goals. Our lean efforts are already producing benefits in the form of increased cash flows, which we are using to repay debt and improve the company's liquidity position."

Lean Boot Camps held throughout the world have helped to improve operational efficiencies and profit margins.

Positioning for the Future

Even given the weak global economy, Gardner Denver remains strong. In just the short span of fifteen years, Gardner Denver has emerged as a leading global manufacturer of highly engineered reciprocating, rotary, and vane compressors; liquid ring pumps and blowers for various industrial and transportation applications; pumps used in the petroleum and industrial markets; and other fluid transfer equipment serving chemical, petroleum, and food industries.

Gardner Denver continues to focus its strategy on growth through acquisition, increased market share, product development, cost reductions, and penetration of niche markets expanding faster than the global economy. The company is organized into two reportable segments: Compressor and Vacuum Products and Fluid Transfer Products.

When Gardner Denver became a stand-alone company in 1993, it was a small compressor, blower, and pump manufacturer, with established operations serving the compressor and industrial blower market. At the time, Gardner Denver facilities were almost completely confined to the United States. During the next fifteen years, the company acquired and opened locations in more than thirty countries worldwide. The company notes that it is strategically positioned to support its customers anywhere in the world. With manufacturing and remanufacturing facilities in the United States, Canada, Brazil, the United Kingdom, Germany, Italy, Finland, China, and Taiwan, the company is truly a global manufacturing powerhouse. Service, distribution, and assembly facilities in Europe, Asia, and North America enable Gardner Denver to respond quickly to customer needs in any time zone anywhere in the world.

Gardner Denver brands are known and trusted everywhere in the world. The Gardner Denver brand remains the largest in the family, but it is supplemented by the Champion, Emco Wheaton, Nash, Elmo Rietschle, and Thomas brands. Gardner Denver continues to make it a policy to preserve brand equity in the world marketplace, while establishing the acquired brands within the Gardner Denver family.

Gardner Denver products on display at the Offshore Technology Conference, the world's largest event that features the latest oil and natural gas discovery and production technology from both onshore and offshore resources

Gardner Denver prides itself on creating and providing innovative compressed air and gas, vacuum, and fluid transfer solutions for customers around the world. The company and its brands are widely recognized as international leaders in technological innovation in many industries, including petroleum, chemical, medical, and wastewater treatment.

Recent technologies used in developing Gardner Denver production include Finite Element Analysis, Computational Fluid Dynamics, Thermal Imaging, Failure Mode Affects Analysis, and Design for Manufacturing. Gardner Denver continues to emphasize research and development as a critical means of helping customers add value to their products. Research and development investment also leads to the near constant introduction of new and improved products that help customers become more efficient at lower costs.

Gardner Denver's inherent strengths position it for continued growth in the global economy of the twenty-first century. A rich heritage stretching back 150 years is a constant reminder to the Gardner Denver family that the challenges of the future are no more insurmountable than those that the company has faced in the past.

Change, challenges, and opportunities have been a way of life for Gardner Denver and its people in the century-and-a-half since Robert W. Gardner set up a small factory just off the Quincy Town Square in 1859. The company changed from a manufacturer of governors for steam boilers to a more diversified firm supplying pumps and compressors to the petroleum industry and drilling equipment to the mining and construction industries. It changed further when it became a subsidiary of Houston-based Cooper Industries in 1979. And the pace of change accelerated when Cooper Industries spun Gardner Denver off as an independent stand-alone company manufacturing compressors, blowers, and pumps.

What hasn't changed is the commitment and dedication of Gardner Denver's global workforce. That's an integral part of the Gardner Denver heritage that Robert W. Gardner and W. H. Leonard would no doubt recognize and applaud.

Gardner Denver Inc. headquarters in Quincy, Illinois

Touching Lives

Most people are totally unaware of how often they come into contact with products they use in everyday life that use technologies provided by Gardner Denver and its growing family of brands. So meet thirty-four-year-old Jennifer from Illinois and find out how Gardner Denver affects her life on a typical day:

6:30 a.m. The electric alarm clock wakes up Jennifer. Gardner Denver pumps help generate electricity in power plants.

6:35 a.m. She gets up and takes a shower. Gardner Denver vacuum pumps and compressors help purify water in many communities.

6:55 a.m. Jennifer brushes her teeth, applies her make-up, and gets dressed. When she goes to see her dentist, she will very likely come upon a suction device with a built-in Gardner Denver vacuum pump. The company's pumps and compressors are also used in the cosmetics and textile industries.

7:10 a.m. The kids are up, and she's preparing breakfast with orange juice, milk, ham, bread, and butter. Cartons for holding liquids are often manufactured using Gardner Denver vacuum technology. Vacuum is applied in many processes in the food and beverage industry—from filtering to packaging. Modern milking systems use vacuum to milk cows. Gardner Denver pumps and blowers also help produce butter and other dairy products.

8:30 a.m. Jennifer scans the newspaper after the kids leave for school. There is virtually no stage in the production of paper that does not use our products: Gardner Denver products remove water from pulp in paper mills and are a vital part of prepress, press and postpress printing applications.

8:45 a.m. Time for work. Jennifer is a hospital administrator. On her way to the hospital, she fills the tank and checks her tire pressure. Robert Gardner started his company in 1859 with products sold to the oil and natural gas industry. Today, Gardner Denver is a leading supplier of equipment for the exploration, production, refining, safe transportation, and distribution of petroleum products. Gardner Denver air compressors and nitrogen generators are found in service stations around the corner and around the world.

9:00 a.m. Arrival at the hospital, where she's in charge of buying medical equipment. Nebulizers, sterilizers, aspirators, air mattresses, and patient-monitoring systems all feature Gardner Denver products.

12:35 p.m. Lunch break. She goes outside and has a soda in the park. Gardner Denver blowers are used in bottle-filling plants. As for producing the plastic material the bottle is made of—Gardner Denver's know-how ranges from degassing to extruding and forming.

4:35 p.m. On her way home from work, she picks up the kids at the bus stop. Miniature pumps and compressors are used for air doors and air suspension. Gardner Denver also provides fleet bus refueling systems.

6:35 p.m. Jennifer's husband Mark returns from a business trip. Aviation is another big industry for Gardner Denver: cabin pressure, breathing air, refueling, tarmac cleaning, and pumping nitrogen into tires to keep them cool are typical applications.

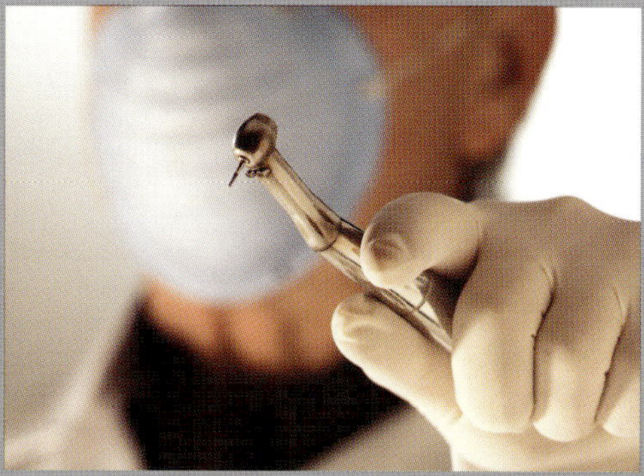

7:35 p.m. Over dinner, she makes plans for the following weekend. The tasks for her husband include arranging for the carpet to be cleaned in the family room and getting the cars washed. Mark is not very enthusiastic. Gardner Denver provides blowers used in the carpet cleaning industry and high-pressure washers used for washing vehicles.

10:35 p.m. Jennifer's to-do list is growing and growing. She wants to redecorate the living room and paint the walls a new color. The swimming pool needs to be cleaned, and the windshield of her car must be replaced. Again, a wide choice of products that use Gardner Denver pumps and compressors are available for these jobs. Chemical plants use Gardner Denver products to produce paint, as do glass manufacturers in their production processes.

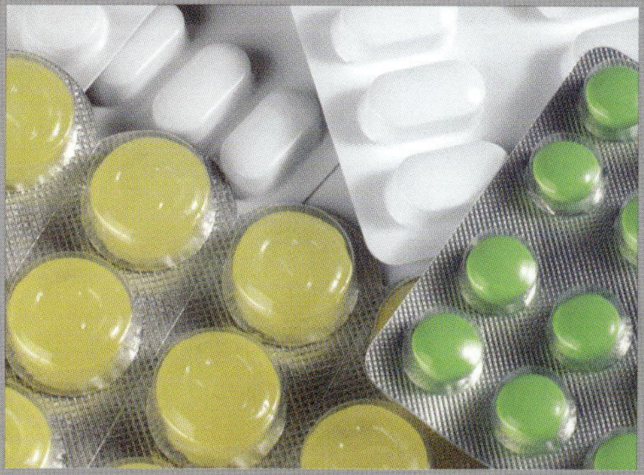

11:35 p.m. One of the children wakes up with a bad cough and needs medication. The pharmaceutical industry uses Gardner Denver pumps and compressors for a wide range of processes, including forming pills, drying, and packaging.

Gardner Denver

INDEX

Page numbers in italics refer to images.